Six Lectures

on Modern Natural Philosophy

Clifford Ambrose

C. Truesdell

Springer-Verlag Berlin Heidelberg New York 1966

Dedicated to my mother

Helen Truesdell Heath

in lifelong gratitude for her efforts to teach me discipline.
method, scholarship, taste, and style.

Foreword

These lectures were first given during my tenure of a Walker-Ames Visiting Professorship in the Department of Astronautics and Aeronautics at the University of Washington, November 2—12, 1964. I am grateful for the interest shown there and for the tranquil hospitality of Dr. JOHN BOLLARD and Dr. ELLIS DILL, which allowed me the leisure sufficient to write the first manuscript.

I thank Dean ROBERT ROY and Dr. GEORGE BENTON for the unusual honor of an invitation to deliver a series of public lectures at my own university.

Apart from the footnotes on pp. 49, 50, and 85, which have been added so as to answer questions allowed by the slower pace of silence, and the obviously necessary note on p. 106, the lectures of this second series are here printed as read, February 9—25, 1965. Thus I may call these, in imitation of a famous example, "Baltimore Lectures".

Acknowledgment

The first lecture is based largely upon my Bingham Medal Address of 1963, part of which it reproduces verbatim. The fifth lecture may be regarded as a partial summary of my course on ergodic theory at the International School of Physics, Varenna, 1960. Much of the last lecture runs parallel to my article "The Modern Spirit in Applied Mathematics", ICSU Review of World Science, Volume 6, pp. 195—205 (1964), and some paragraphs are taken from my address to the Fourth U.S. National Congress of Applied Mechanics (1961). For permission to reprint the passages in question I am grateful to the editors of the volumes, to Interscience Publishers, to the Elsevier Publishing Company, and to the American Society of Mechanical Engineers.

To the U.S. National Science Foundation is owed double gratitude: first, for partial support of much of the research by the various savants who created the new science these lectures resume, and, second, for partial support of the writing of the lectures themselves.

For their kind criticism of the manuscript I am grateful to Messrs. COLEMAN, ERICKSEN, NOLL, and TOUPIN.

Contents

I. Rational mechanics of materials

In all of natural philosophy, the most deeply and repeatedly studied part, next to pure geometry, is mechanics. The resurgence of rational mechanics, after half a century's drowsing, has signalled and led the rediscovery of natural philosophy as a whole, just beginning in our time. For two hundred years, the fields of scientific research were wilfully shrunk and sharpened to pin-point size, and appropriate microscopes were developed so that organized micro-thought could split them into forklets of microscience, now budgeted at rates in megabucks per kilohour. Opening one's eyes to the daylight after having been trained to distrust anything seen without squinting is not easy. As during RIP VAN WINKLE's dream, the perceptible spectrum of nature has changed while we crept into ever deeper and darker corners, but it has not narrowed. To see it again, we need the telescope of generality, but looking through the wrong end of the microscope of ever finer subdivision will give us a worse picture of nature than was seen by the alert, unaided eye in the seventeenth century or may be learned today from the comic strips called "science for the layman".

For making a telescope, experience in the design of microscopes, while not sufficient, is also not useless. We do not wish to join the theologians in uncritical awe before a merely devotional image of nature, or to float with the professional philosophers as they stamp in a quicksand while jotting down their queries. We wish an instrument of precision. It is fitting to gain experience with an objective of modest power but modern design.

Dropping the metaphor, I turn to a theory of modest generality, which is just what mechanics is. The picture of nature as a whole given us by mechanics may be compared to a black-and-white photograph: It neglects a great deal, but within its limitations, it can be highly precise. Developing sharper and more flexible black-and-white photography has not attained pictures in color or three-dimensional casts, but it serves in cases where color and thickness are irrelevant, presently impossible to get in the required precision, or distractive from the true content. The new rational mechanics serves the same function with regard to natural philosophy as a whole. If I dared, I might even throw a parallel to the arts, where some of the greatest painters and sculptors have for certain subjects set their palettes or hammers aside and chosen the pen or the burin as fitter than the tinted brush or the gouging chisel. No one has suggested doing REMBRANDT'S etchings over again in color or relief.

The science we call to our aid in constructing instruments with which to see nature is mathematics. Rational mechanics was the first domain of natural philosophy on which modern mathematics was brought to bear so as to form a real theory, comparable in generality and precision to classical geometry. In this lecture I shall outline the simplest part of the rational mechanics of materials, as organized by Mr. NOLL.

The following three concepts, direct idealizations from daily experience, are the primitive elements of mechanics.

1. The concept of a body, $\mathscr{B}$. A body is a three-dimensional smooth manifold, the elements of which are *particles* **X**. Since experience does not determine any particular geometry in bodies, it would be presumptuous to enforce one. Geometrical structure should be dictated by physics, not vice versa. The little experience common to all gross bodies is summed up in two statements. (A) Every sufficiently smooth portion of a body is a body, and a measure, called *mass*, is defined over the body. (B) A body is never accessible directly to observation, since we encounter it only at particular times and particular places. These times and places form the space-time manifold of physical experience. In "classical" mechanics we postulate that this manifold is Euclidean and three-dimensional, thus employing the second primitive element:

2. Euclidean space-time, consisting of points **x**, endowed with the Euclidean metric structure, which is independent of the time, t. In continuum mechanics, a body $\mathscr{B}$ may be mapped smoothly onto regions of Euclidean space. The region occupied by the body is its *configuration.* Only in some configuration is a body ever liable to be observed. The *motion* of a body is the specification of the places taken by its particles as time goes on, or the time sequence of its configurations:

$$\boldsymbol{x} = \boldsymbol{\chi}_{\raisebox{1pt}{.}}(X) = \boldsymbol{\chi}(X, t),$$

where $\boldsymbol{\chi}_t$, for each t, is a continuously differentiable homeomorphism.

3. The third element of mechanics is a *system of forces*. Continuum mechanics concerns contact forces, which express the reaction of material upon material in a given configuration of a body. The stress vector $\boldsymbol{t}_{(\boldsymbol{n})}$ acting upon a boundary with outer normal $\boldsymbol{n}$ is gotten from the stress tensor $\boldsymbol{T}$:

$$\boldsymbol{t}_{(\boldsymbol{n})} = \boldsymbol{T}\boldsymbol{n}.$$

Mechanics seeks to connect these three elements — body, motion, and force — in such a way as to yield good models for the behavior of the materials in nature. There are two steps.

First, we postulate principles *common to all bodies*, namely, the conservation laws for force and torque, or in other terms, for linear momentum and moment of momentum. When the fields occurring are sufficiently smooth, these principles assume the well known form of CAUCHY's laws of motion:

$$\operatorname{div}\boldsymbol{T} + \varrho\,\boldsymbol{b} = \varrho\,\ddot{\boldsymbol{x}}, \qquad \boldsymbol{T} = \boldsymbol{T}^{\prime},$$

where ϱ is the mass density, $\boldsymbol{b}$ is the body force, $\ddot{\boldsymbol{x}}$ is the acceleration, and superscript T denotes transposition. These principles are satisfied by all bodies, independently of their material properties.

Experience shows us that two bodies of different material generally behave differently when subjected to the same external forces. Therefore, the internal forces must be different in the two cases, since if both body force $\boldsymbol{b}$ and stress tensor $\boldsymbol{T}$ are given, a unique value for the acceleration $\ddot{\boldsymbol{x}}$ follows from CAUCHY's first law. Conversely, if $\boldsymbol{b}$ and $\ddot{\boldsymbol{x}}$ are given, CAUCHY's first law gives us a unique value for div $\boldsymbol{T}$, not for $\boldsymbol{T}$. That is, the general principles of mechanics do not suffice to determine the forces acting within a body subjected to a given motion.

The second part of the task of mechanics is to represent the *variety* of materials. To this end we frame *constitutive equations*, which determine the stress from sufficient data of motion. What

we know to be true and what we believe to be reasonable for one
or another real material serve as our guides in choosing different
forms of constitutive equations.

A *general theory of constitutive equations* may be founded upon
the following three principles:

1. Principle of determinism. The stress in a body is determined
by the history of the motion of the body:

$$\boldsymbol{T}(X, t) = \mathfrak{F}(\boldsymbol{\chi}; X, t).$$

Here $\mathfrak{F}$ is a functional of the motion $\boldsymbol{\chi}$ of the body $\mathscr{B}$. While some
scientists may not yet be accustomed to the concept of "function-
al", it is simpler and much closer to physical experience than are
"derivative" and "integral", not to mention that ubiquitous
weasel, "approximation".

A *functional* of the function f is a quantity determined by f,
nothing more, nothing less. For example, the area under the
curve $y = f(x)$ from $x = a$ to $x = b$ is a functional of f, depending
upon a and b as parameters. While in earlier work only special
kinds of functionals are used, here the term "functional" is taken
in its broadest and simplest sense. By "the motion $\boldsymbol{\chi}$" is here
meant the *restriction* of $\boldsymbol{\chi}(X, \tau)$ to the range $-\infty < \tau \leqq t$, where
t is the present time. Thus the general constitutive equation says
neither more nor less than the following sentence: Knowledge of
the motion of all particles of the body at the present and all past
times suffices to determine the present stress at the particle X.
Note that only *past* experience is relevant. The widespread belief
that mechanics concerns only processes reversible in time is a
pure illusion, not worth the trouble of refuting. $\mathfrak{F}$, the *constitutive
functional* defining the particular material, expresses all the information we have about the mechanical properties of the particles
X which make up the body.

2. Principle of local action. According to the principle of
determinism, the stress at the particle X could be influenced by
the experiences of a distant particle Z. This is surely not correct.
Stress forces are contact forces and should be determined by
contact actions alone, even if, perhaps, contact actions of long ago.
Accordingly, we impose upon the functional $\mathfrak{F}$ a restriction forbid-

ding action at a distance in material response. Namely, if two motions χ and $\bar{\chi}$ differ from one another only outside of an arbitrarily small neighborhood $\mathcal{N}(X)$ of X, for all times $\tau \leq t$, then they give rise to the same present stress at X. Formally, if $\chi(Z, \tau) = \bar{\chi}(Z, \tau)$ for $\tau \leq t$ and for $Z \in \mathcal{N}(X)$, then

$$\mathfrak{F}(\chi; X, t) = \mathfrak{F}(\bar{\chi}; X, t).$$

3. Principle of material frame-indifference. If two observers consider the same motion in a given body, they find the same state of stress. An "observer" in classical mechanics is a rigid frame which bears a clock. If $\chi(X, t)$ is the motion as apparent to the first observer, and if $\chi'(X, t')$ is the motion as apparent to the second observer, it can be proved that

$$\chi' = Q \chi + c, \quad t' = t - a,$$

where Q is a time-dependent orthogonal tensor, c is a time-dependent vector, and a is a constant. According to the principle of determinism, the second observer finds the stress T' from the motion as follows:

$$T'(X, t') = \mathfrak{F}(\chi'; X, t'),$$

where $\mathfrak{F}$ is the constitutive functional, which is laid down once and for all in definition of a particular material. According to the principle of material frame-indifference, T' and T yield the same stress vectors at every point and for every surface element; that is,

$$T' = Q T Q^T.$$

Substituting for T' and T their expressions in terms of $\mathfrak{F}$ yields a *functional equation* to be satisfied by every constitutive functional:

$$\mathfrak{F}(Q \chi + c; X, t - a) = Q \mathfrak{F}(\chi; X, t) Q^T,$$

identically in Q, χ, c, and a.

The principle of material indifference may be rendered plausible in several ways. For one thing, it is satisfied by every classical theory of materials, and also by most of the less familiar proposals of constitutive equations. For another, without such a principle a number of generally accepted ideas in physics would become meaningless. Consider, for example, what are called centrifugal forces. Take a spring, and on one end hang a weight of one pound. The spring lengthens, say by one inch. Now lay the spring on a horizontal table, fastening one end to the center, and leaving the weight attached to the other end. Spin the table, and adjust the angular speed until the spring again stretches exactly one inch. On seeing this demonstration in the laboratory, the freshmen, happy to participate in the experimental foundation of science, take it as obvious that the force exerted by the spring is again one poundal; substituting the appropriate numbers into the formula for centrifugal force, they are encouraged to join GALILEO and MACH and LEONARDO DA VINCI as empiricists by calculating an angular speed that agrees with the measured value, to within experimental error. What has been assumed, tacitly, is that the elastic law or constitutive equation of the spring is invariant under rotation. For an observer standing on the floor as well as for an observer seated upon the table, or, for that matter, for an observer watching the experiment in a plane mirror as he is shot from the mouth of a cannon, one inch of extension corresponds to one poundal of force. To a person lacking this belief, the experiment measures nothing. Such a person sees two events but can assert no correlation of one with another. (As is usual in "fundamental" experiments, the main point at issue must be conceded before the experiment is started.)

If not everyone is content with this explanation, it may be some solace to mention that ten years ago I was not convinced that the principle of material indifference is general and unexceptionable. Today, albeit with some difference of form, it is already thrown at beginners in the all too popular engineering science of theory-creating as being "obvious" and "intuitive".

The three fundamental principles may be repeated in words:

1. *Determinism.* The stress in a body is determined by the motion the body has undergone.

2. *Local action.* The motion outside an arbitrarily small neighborhood of a particle may be disregarded in determining the stress at that particle.

3. *Material indifference.* Any two observers of a motion of a body find the same stress.

The equations given above mean just the same thing as these statements in words.

The theory based on these principles, while by no means general enough to include everything presently studied in continuum mechanics, allows for material behavior of great complexity. An important special case is defined by requiring material response to be determined for *all* deformation processes as soon as it is known for all *homogeneous* deformation processes. A material of this kind is called *simple*. To define a homogeneous deformation, consider a particular configuration $\varkappa$ of a body $\mathscr{B}$. The equation

$$\boldsymbol{X} = \varkappa(X)$$

states that the particle X occupies the place $\boldsymbol{X}$ in the configuration $\varkappa$. The place $\boldsymbol{X}$ is the "material co-ordinate" of the particle X with respect to the configuration $\varkappa$. As the same point in geometry has infinitely many different co-ordinates, according to which co-ordinate system we may choose to employ, so the particle X has infinitely many material co-ordinates $\boldsymbol{X}$, one for each configuration $\varkappa$, and $\varkappa$, like a co-ordinate system, is entirely at our disposal. Since $\varkappa$ is an invertible mapping, the motion χ may be written with $\boldsymbol{X}$ as independent variable:

$$\boldsymbol{x} = \chi(X, t) = \chi\left(\varkappa^{-1}(\boldsymbol{X}), t\right) \equiv \chi_{\varkappa}(\boldsymbol{X}, t),$$

say. Then

$$\boldsymbol{F}(\boldsymbol{X}, \tau) \equiv \nabla \chi_{\varkappa} \equiv \frac{\partial \chi_{\varkappa}(\boldsymbol{X}, \tau)}{\partial \boldsymbol{X}}$$

is the gradient of the deformation from $\varkappa$ to $\chi_{\varkappa}$. ($\boldsymbol{F}$ is always invertible.) We are interested in the *deformation history* at $\boldsymbol{X}$, namely, the totality of values of $\boldsymbol{F}(\boldsymbol{X}, \tau)$ for all times τ up to the present time t. A deformation history is called *homogeneous* when $\boldsymbol{F}$ is constant in space, at each time. According to our definition, to determine the stress in a simple material all we need know about the motion $\chi_{\varkappa}$ is the history of its gradient $\boldsymbol{F}(\tau)$, $-\infty < \tau \leq t$. That is, the *constitutive equation of a simple material* has the form

$$\boldsymbol{T} = \mathop{\mathfrak{G}}_{\tau=-\infty}^{\tau=t} \left(\boldsymbol{F}(\tau)\right),$$

where $\mathfrak{G}$, the constitutive functional, may depend upon $\varkappa$, $\boldsymbol{X}$, and t as parameters. (We shall not write the limits $-\infty$ and t except when we wish to emphasize them.) While the definition itself seems to depend upon the choice of reference configuration, such dependence is easily proved illusory.

The constitutive equation of a simple material satisfies trivially the principles of determinism and local action. To consider the principle of material frame-indifference, we notice that under change of observer $\boldsymbol{F}$ becomes $\boldsymbol{F}'$, where

$$\boldsymbol{F}' = \boldsymbol{Q}\,\boldsymbol{F}.$$

The functional equation expressing the principle of material frame-indifference becomes

$$\mathfrak{G}\big(\boldsymbol{Q}\,(\tau)\,\boldsymbol{F}\,(\tau)\big) = \boldsymbol{Q}\,(t)\,\mathfrak{G}\big(\boldsymbol{F}\,(\tau)\big)\,\boldsymbol{Q}\,(t)^T,$$

identically in $\boldsymbol{Q}$ and $\boldsymbol{F}$. Since $\boldsymbol{F}$ can be split into a pure deformation $\boldsymbol{U}$ followed by a rotation $\boldsymbol{R}$, that is,

$$\boldsymbol{F}\,(\tau) = \boldsymbol{R}\,(\tau)\,\boldsymbol{U}\,(\tau),$$

the functional equation may be written as

$$\mathfrak{G}\,[\boldsymbol{Q}\,(\tau)\,\boldsymbol{R}\,(\tau)\,\boldsymbol{U}\,(\tau)] = \boldsymbol{Q}\,(t)\,\mathfrak{G}\,[\boldsymbol{F}\,(\tau)]\,\boldsymbol{Q}\,(t)^T.$$

For $\boldsymbol{Q}\,(\tau)$ choose the special rotation $\boldsymbol{R}\,(\tau)^T$. The rotation then cancels out on the left-hand side:

$$\mathfrak{G}\big(\boldsymbol{U}\,(\tau)\big) = \boldsymbol{R}\,(t)^T\,\mathfrak{G}\big(\boldsymbol{F}\,(\tau)\big)\,\boldsymbol{R}\,(t).$$

Therefore the constitutive equation of a simple material may be written in the following *reduced form:*

$$\boldsymbol{T} = \boldsymbol{R}\,(t)\,\mathop{\mathfrak{G}}_{\tau=-\infty}^{\tau=t}\,[\boldsymbol{U}\,(\tau)]\,\boldsymbol{R}\,(t)^T,$$

by which the functional equation expressing the principle of material frame-indifference is solved, once and for all, for simple materials. The stress tensor may depend in any way on whatever pure deformation the particle has experienced; past rotations have no effect on the present stress; and the role of the present rotation is made explicit and shown to be exactly the same as in the classical theory of finite elastic strain.

Up to now we have considered a single particle, X, whose position in the reference configuration $\varkappa$ is denoted by X. Now consider a second particle, $\widehat{X}$. When can we say that X and $\widehat{X}$ are of the same material? When it is possible to bring the portions of the body near to the two particles X and $\widehat{X}$ into configurations of equal and uniform density such that, starting from these, any particular subsequent deformation process gives rise to exactly the same stress for each of the particles. In mathematical terms, X and $\widehat{X}$ are *materially isomorphic* if there exist reference configurations $\varkappa$ and $\widehat{\varkappa}$ such that

$$\varrho_{\varkappa}(X) = \varrho_{\widehat{\varkappa}}(\widehat{X}) = \text{const.},$$

$$\mathfrak{G}_{\varkappa}[F(\tau);\, X] = \mathfrak{G}_{\widehat{\varkappa}}[F(\tau);\, \widehat{X}],$$

for every history $F(\tau)$. The second equation states that the stress at X arising from the deformation history F is just the same as that at $\widehat{X}$ for the same deformation history. So as to emphasize the dependence of the constitutive functional upon the reference configuration, we have written $\mathfrak{G}_{\varkappa}$ and $\mathfrak{G}_{\widehat{\varkappa}}$ instead of $\mathfrak{G}$. We can repeat the definition in other words: The particles X and $\widehat{X}$ are materially isomorphic if no experiment can tell us whether we are confronted with X after a deformation starting from its place X in $\varkappa$ or with $\widehat{X}$ after a deformation starting from its place $\widehat{X}$ in $\widehat{\varkappa}$.

It is possible that a particle may be materially isomorphic to itself in a non-trivial way. That is, there may exist two different configurations, $\varkappa$ and $\widehat{\varkappa}$, such that the response of the particle X after deformation from its place in either of these is just the same. Indeed, let H be a unimodular tensor — one whose determinant is ± 1 — such that

$$\mathfrak{G}\big(F(\tau)\big) = \mathfrak{G}\big(F(\tau)\,H\big).$$

By the chain rule in calculus, H^{-1} is the gradient at X of the mapping from $\varkappa$ to a new reference configuration, $\widehat{\varkappa}$, for the same particle X, such that no experiment on that particle can determine whether we began with it in $\varkappa$ or in $\widehat{\varkappa}$. That is, the two configurations $\varkappa$ and $\widehat{\varkappa}$ are physically indistinguishable, as far as the particle X is concerned. The set of all such H forms a group, g, called the *isotropy group* of the material at the particle X, with respect to the reference configuration $\varkappa$. The elements of this group generate

all deformations of a neighborhood of X in $\varkappa$ which are indistinguishable by experiment; that is, no experiment can decide whether
a deformation generated by the isotropy group has occurred or not.

The transformations $\boldsymbol{H}$ in the isotropy group g need not be
orthogonal, but they may be. The orthogonal members of g are
those orthogonal tensors $\boldsymbol{Q}$ that are solutions of the equation

$$\mathfrak{G}\,(\boldsymbol{Q}\,\boldsymbol{F}\,\boldsymbol{Q}^{T}) = \boldsymbol{Q}\,\mathfrak{G}\,(\boldsymbol{F})\,\boldsymbol{Q}^{T}$$

for all $\boldsymbol{F}$, as follows from the previous equation by the principle
of material frame-indifference. Hence, in particular, the identity $\mathbf{1}$
and the central inversion $-\mathbf{1}$ belong to every isotropy group.

Thus far the reference configuration $\varkappa$ has been fixed. To
emphasize the dependence of g on $\varkappa$ we may denote it by $g_{\varkappa}$. If we
begin with a different reference configuration $\hat{\varkappa}$, we obtain, in
general, a different isotropy group, $g_{\hat{\varkappa}}$. Let $\boldsymbol{K}$ be the gradient of
the deformation from $\varkappa$ to $\hat{\varkappa}$. Then it is easy to prove NOLL's rule:

$$g_{\hat{\varkappa}} = \boldsymbol{K}\,g_{\varkappa}\,\boldsymbol{K}^{-1}.$$

Therefore, if the isotropy group for one configuration is known,
we can determine it at once for all other reference configurations.
Many of the physical properties of a simple material may be
specified in terms of the isotropy groups of its particles.

It used to be said that a fluid is a material that flows. That is,
a fluid body cannot remain in equilibrium when subject to shear
stress upon its boundaries. We know now, both by theory and by
experience, that also some solids can flow. Likewise, while a
fluid is known to have no preferred configurations:

$$g_{\varkappa} = g_{\hat{\varkappa}} \quad \text{for all configurations } \varkappa,\,\hat{\varkappa},$$

this property also is not peculiar to fluids, since it is shared by
materials having the smallest possible isotropy group, $g = \{\mathbf{1},\,-\mathbf{1}\}$.
In such a material, called "triclinic", *every* non-trivial deformation
from *any* configuration produces a configuration with different
physical response. Thus lack of preferred configurations can mean
simply that the response of the material is so complicated as to
show no symmetries in any configuration.

A fluid has properties of the opposite kind. It has a vast
number of symmetries, and it has them in every configuration.
From its definition, the isotropy group of any material is a sub-

group of the unimodular group:

$$g < u.$$

We define a *simple fluid* as a material having the greatest possible material symmetry; namely, a simple material whose isotropy group is the full unimodular group:

$$g = u.$$

Again, obviously, $g_\varkappa = g_{\hat\varkappa}$ for all $\varkappa$ and $\hat\varkappa$, so that a fluid shares with a triclinic material the property of having no preferred configurations, but for the opposite reason, namely, that every configuration shows the maximum possible symmetry of response. It can be shown that fluids and triclinic materials are the only materials having an isotropy group the same for all reference configurations.

The concept of "simple fluid", being a direct statement of invariance, is not to be confused with older, more formal definitions fashioned by extraneous and restrictive mathematical apparatus such as derivatives, integrals, and polynomials. It includes as special cases Reiner-Rivlin fluids, Rivlin-Ericksen fluids, some of OLDROYD's fluids, and various other recent proposals but allows in addition the possibility of manifold effects of stress relaxation. Let me repeat the definition in other words: A simple fluid is a material whose entire mechanical behavior may be determined from experiments on homogeneous strain and whose response to a deformation from one configuration with uniform density is the same as from any other with the same uniform density.

This fact is expressed in mathematical form through a fundamental theorem of Mr. NOLL. Let $G_t(s)$ be the strain tensor for the configuration at time $t - s$ when the *present* configuration, that at time t, is taken as reference. Specifically, $G_t(s) + 1$ is the right Cauchy-Green tensor, with respect to the configuration at time t, for the configuration at time $t - s$. Then a material is a simple fluid if and only if

$$T = -p\,1 + \overset{\infty}{\underset{s=0}{\mathfrak{G}}}\,(G_t(s);\,\varrho), \quad \mathfrak{G}(0;\,\varrho) = 0,$$

where p is the equilibrium pressure. As was to be expected, a fluid turns out to be a substance capable of flow, since its stress in equilibrium must reduce to a hydrostatic pressure. The non-equilibrium part of the constitutive equation delimits the peculiar kind of memory a simple fluid may have. A simple fluid may remember everything that ever happened to it, yet it cannot recall any one configuration as being physically different from any other. It reconciles these two almost contradictory qualities by remembering the past only in comparison with the ever-changing present. $G_t(s)$ measures the backward strain, the strain of the past configurations with respect to the present one, and it is only through $G_t(s)$ that the fluid responds to its past.

A solid has very different properties. It has preferred configurations, with respect to which its response appears particularly simple. For example, if a solid bar is stretched and then held fast in its lengthened state, some force must continue to be supplied, at least for a long time, and, whatever be the force required to effect this elongation, there is no reason to expect that this same force will effect the same elongation again if we start from this new configuration of the bar. The response of the material in the deformed state will be different from its response in the original or undistorted state. For a solid, then, there must be at least one configuration from which any non-orthogonal deformation brings the body into a configuration with different response. Formally, a simple material is a *simple solid* if there exists at least one configuration, called an *undistorted state*, such that

$$g \subset o ,$$

where o is the orthogonal group.

A material is *isotropic* if it has at least one configuration, again called an *undistorted state*, every orthogonal deformation from which carries the body into a physically indistinguishable configuration. That is, a material is isotropic if and only if it can be brought into a configuration from which rotations cannot be detected by experiment. Formally,

$$g \supset o .$$

From these definitions, which, being direct translations into mathematics of simple and clear physical concepts, now seem obvious, follow some immediate but not unimportant theorems:

1. Every simple fluid is isotropic.

2. For an isotropic simple solid in an undistorted state, $g = o$.

3. Every isotropic simple material is either a solid or a fluid.

This last fact is only a translation into the present terms of the theorem that the orthogonal group is maximal in the unimodular group.

We can diagram the kinds of groups mentioned so far:

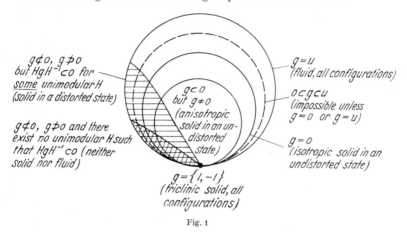

$g \not\subset o$, $g \not\supset o$ but $HgH^{-1} \subset o$ for *some* unimodular H (solid in a distorted state)

$g \not\subset o$, $g \not\supset o$ and there exist no unimodular H such that $HgH^{-1} \subset o$ (neither solid nor fluid)

$g \subset o$ but $g \neq o$ (anisotropic solid in an undistorted state)

$g = \{1, -1\}$ (triclinic solid, all configurations)

$g = u$ (fluid, all configurations)

$o \subset g \subset u$ (impossible unless $g = o$ or $g = u$)

$g = o$ (isotropic solid in an undistorted state)

Fig. 1

Now there are some subgroups of the unimodular group which neither contain the orthogonal group nor are contained by it. Let g be such a group. Trivially, it cannot correspond to a fluid. If there exists a unimodular tensor H such that $HgH^{-1} \subset o$, NOLL'S rule shows that the material is a solid, but that the reference configuration is not undistorted. If there exists no such H, the material fails to satisfy the definition of a solid. It is known that there are groups of this kind. Therefore,

4. There are simple materials that are neither solids nor fluids. Such a material is one for which, from every configuration, there exists at least one non-orthogonal deformation under which the physical properties of the material remain unchanged, yet there is some density-preserving deformation that does alter its response.

While, of course, the existence of subgroups of the unimodular group that are not conjugate to any orthogonal subgroup has long been known, in the last few weeks Mr. WANG in his dissertation, just accepted by this university, has distinguished and interpreted a class which defines simple materials whose symmetries are suggested by those of various kinds of liquid crystals. Theories to represent these physical materials had been proposed earlier, within a different framework, by Mr. ERICKSEN. WANG'S work shows that NOLL'S theory of simple materials is broad enough to include response midway, in various senses, between the common ideas of "fluid" and "solid". WANG'S *subfluids*, as he calls them, retain their properties under dilatations along a preferred triad of axes, but, in general, under no other changes of shape. Subfluids can support in equilibrium certain kinds of shear stress, but will flow if sheared in other directions.

WANG has proved that there are exactly fourteen types of subfluids and has determined their symmetries and other distinguishing properties.

I said at the outset that NOLL'S theory of materials is simple and easy. "Simple and easy" does not and should not mean "expressed solely in terms of mathematics created 200 years ago and now taught to science students as an article of religion." The main mathematical concepts used here are: manifold, smooth mapping, Euclidean space, vector, tensor, functional, and group. In each case, little more than the definition, or the concept itself, has been needed; the grinding operations commonly called "mathematics" by physical scientists are of no use here. While more recently formalized, the concepts employed in NOLL'S theory are closer to common, untrained experience than are such things as Fourier coefficients, Laplace transforms, complex variables, Bessel functions, polynomial bases, *etc.*, which many scientists seem to find more familiar. Exclusive mathematical development of linear theories in the last century formed specialized mathematical tools that lie far away from physical experience but come to mind at once because they have been learned in school. Just as some experimental apparatus grows obsolete and can be forgotten by the modern experimenter, much of the mathematical apparatus of our grandfathers' time, especially that which physicists and engineers usually insist is essential "applied mathematics", is in fact poor dull stuff the theorist today can safely cast aside in favor of the sharper tools of modern algebra and analysis. NEWTON said, "Nature is simple, and affects not the pomp of superfluous causes." To deal with general response we must learn to think simply again, to use mathematical concepts that represent experience unblunted and unblurred.

Not long ago was born a new branch of continuum mechanics called the "theory of dislocations". Since this theory has grown up mainly in laboratories of solid-state physics, it has been presented in a language of its own, in which clarity and logic take a poor second place to physical intuition. What is assumed is rarely distinguished from what is left to be proved, and the final appeal, in every case of doubt, seems to be to a model of matter as composed of small, hard balls. On this basis, a peculiar geometric structure, not occurring elsewhere in mathematical physics, is assigned *a priori* to the body and is discussed in the unnumbered repetitions of an exuberant literature. Those who have failed to become converts to the hard-ball view are told they lack physical intuition. This is surely true but does not obviate the need for a clear, open, logically sound, continuum basis for a continuum theory. The history of physics shows that different, apparently almost contradictory hypotheses of structure and definitions of gross variables based upon them lead to the same equations for continua. From incorrect molecular hypotheses the first correct theories of viscosity in fluids and elasticity in solids were derived. The main worth of a field theory is lost if the results are tied irrevocably to conjectures about particle structure. Molecular hypotheses have come and gone, but a sound continuum theory is a monument forever, exempt from fashion.

One often hears, particularly in comments after lectures, the claim that by introducing the deformation gradient, $\boldsymbol{F}$, the rational mechanics of continua excludes the theory of dislocations at the outset. This misunderstanding may be explained to some extent by the lack of any treatment of dislocation theory in the literature of rational mechanics. In work done two years ago during his tenure of a visiting professorship at this university, but still, unfortunately, unpublished, Mr. NOLL has proved that the theory of dislocations, insofar as it has yet been clothed in concrete mathematical form as distinct from mere picture-drawing, is a special case of his theory of simple materials, obviating the need for physical intuition or other special pleading.

To outline his results, let us return to the definition of material isomorphism. If each particle of a body is materially isomorphic to every other, all particles have exactly the same physical properties, and the body is called *materially uniform*. Recall that to exhibit the isomorphism of X and $\hat{X}$, two different configurations, $\varkappa$ and $\hat{\varkappa}$, were introduced. It is possible, as a special case, that $\varkappa = \hat{\varkappa}$. If a single reference configuration suffices for the isomorphism of all particles of a materially uniform body, then the *whole body* may be brought into a configuration from which each particle responds in just the same way to any given deformation. Such a body is *homogeneous*. In an inhomogeneous but materially uniform body, infinitely many different reference configurations, one for each particle, may be required to verify that all particles do, indeed, exhibit the same response. In this case it is not possible to bring X and $\hat{X}$ *simultaneously* into configurations from which the subsequent response is the same, unless, of course we break up the body into smaller portions. A theory of "dislocations" results necessarily as soon as one considers bodies materially uniform but not homogeneous.

Specifically, the statement that every particle is materially isomorphic to every other is an assertion that a distant parallelism is defined over the body $\mathscr{B}$. The affine connection, $\boldsymbol{\varGamma}$, of this parallelism has vanishing curvature but need not be symmetric. Mr. NOLL has proved the following theorem: A materially uniform body is homogeneous if and only if $\boldsymbol{\varGamma}$ is symmetric. Accordingly, the torsion tensor or antisymmetric part of $\boldsymbol{\varGamma}$,

$$S_{\mu\lambda}{}^{\varkappa} \equiv \varGamma_{[\mu\lambda]}{}^{\varkappa},$$

is called the *inhomogeneity of the body*. Thus it is *proved* that the geometric structure postulated by dislocation experts from hardball motivation must exist in any materially uniform but not homogeneous simple body. "Dislocations" thus appear as a particular kind of inhomogeneity, fixed in the material. Everything follows, as in other parts of continuum mechanics, straight from the constitutive equations for the particles making up the body. So far, these equations need not be specified further: The particles of the body may be solid or fluid or neither, so long as they be all alike. As soon as the physical properties of the body are laid down, the geometrical structure is found, as it should be, by mathematical process.

The dislocation experts, with boundless thirst for finding complications within complications until all hope of ever proving anything is lost before the terrifying complexity of their orgies of formalism, think that the stress should not be symmetric; nay more, now they begin to see bodies filled with multiform dislocations and the corresponding multipolar stresses of all orders. The results of NOLL show that in the theory of simple materials, built up by use of a symmetric stress tensor and all the other concepts of continuum mechanics in their *simplest forms,* the essential structure of dislocation theory necessarily exists. They do not show, of course, that in a more general theory it will not exist; rather, we expect that a similar treatment will be possible for polar media also, but it is only sound economy to explore first the simpler case just given, not adding the complication of polar response until proved necessary. Still more complicated would be a theory of the origin and movement of dislocations within a body. When the dislocationists claim that their theories already describe moving dislocations, we must understand that "theory" for them is a word of looser meaning than in mathematical science. After inferring a geometric structure for the body manifold, the literature of dislocation theory stops short of constitutive equations and turns to the interpretation of photographs and the drawing of suggestive diagrams. Since NOLL's theory is based on constitutive equations from the start, a full set of field equations follows from it, as we shall now mention in a special case.

A material is *elastic* if its constitutive equation is of the form

$$\boldsymbol{T} = \mathfrak{h}(\boldsymbol{F}, \boldsymbol{X}),$$

where $\mathfrak{h}$ is a function and where $\boldsymbol{F}$ is the gradient of the deformation from some reference configuration. Substitution into CAUCHY's first law of motion yields the differential equation

$$\mathrm{Div}\,\mathfrak{h}(\boldsymbol{F}, \boldsymbol{X}) + \varrho_{\mathrm{R}}\boldsymbol{b} = \varrho_{\mathrm{R}}\ddot{\boldsymbol{x}},$$

where ϱ_{R} is the density in the reference configuration. To render this equation explicit, set

$$A_{k\;m}^{\alpha\;\beta} \equiv \frac{\partial \mathfrak{h}_k^{\alpha}}{\partial F^m_{\;\beta}}, \qquad q_k \equiv \frac{\partial \mathfrak{h}_k^{\alpha}}{\partial X^{\alpha}}.$$

Then
$$A_{k\ m}^{\ \alpha\ \beta} F^m_{\ \alpha;\beta} + q_k + \varrho_R b_k = \varrho_R \ddot{x}_k,$$
where
$$F^m_{\ \alpha;\beta} = \frac{\partial^2 x^m}{\partial X^\alpha \partial X^\beta} + \left\{ \begin{matrix} m \\ q\ r \end{matrix} \right\} x^q_{,\alpha} x^r_{,\beta} - \left\{ \begin{matrix} \sigma \\ \alpha\ \beta \end{matrix} \right\} x^m_{,\sigma},$$

the curly brackets being Christoffel symbols based on the Euclidean metric tensor in the co-ordinates selected in the present configuration and in the arbitrary reference configuration, respectively. We may, if we like, choose a Cartesian frame in each, and then the components $F^m_{\ \alpha;\beta}$ reduce to the second derivatives of the deformation:
$$F^m_{\ \alpha;\beta} = \frac{\partial^2 x^m}{\partial X^\alpha \partial X^\beta}.$$

The above differential equation governs the deformation in an elastic material of arbitrary symmetry, with any kind of inhomogeneity, and referred to any reference configuration. Since an equation of this form is always valid, the dislocation experts mislead us when they claim that a deformation does not exist for a body with dislocations. Of course, there is always a deformation, and we may always refer it, without prejudice to the nature of the material, to any reference configuration we please. What the dislocation experts are trying to tell us is that q, the resultant force of inhomogeneity, will not be zero if dislocations are present, and worse than that, we shall not know what q is. Here they are right. The equation is true but useless in an elastic material with dislocations. Since we do not know q, we do not have a definite equation to solve. For a homogeneous material, it is possible to choose a reference configuration such that the constitutive equation has the same form for each particle:
$$T = \mathfrak{h}(F),$$

and hence $q = 0$, whence follows a simple and manageable equation for the displacement, known for over a century:
$$A_{k\ m}^{\ \alpha\ \beta} F^m_{\ \alpha;\beta} + \varrho_R b_k = \varrho_R \ddot{x}_k.$$

What has long been wanted in the theory of elastic dislocations is a similarly simple, explicit equation of motion, an equation exploiting the particular kind of inhomogeneity that corresponds to dislocations.

Mr. NOLL has found this equation. He has shown that for any materially uniform elastic body there exists a "uniform reference" such that if the deformation $\boldsymbol{x} = \boldsymbol{\chi}(\boldsymbol{X}, t)$ is referred to it, a constitutive equation of the form

$$\boldsymbol{T} = \mathfrak{h}(\boldsymbol{F})$$

holds, but the force of inhomogeneity is not generally zero:

$$q_k = -\mathfrak{h}_k{}^\alpha S_{\alpha\sigma}{}^\sigma,$$

where $\boldsymbol{S}$ is the inhomogeneity tensor calculated from $\boldsymbol{\Gamma}$, the affine connection of the distant parallelism. The uniform reference is not a Euclidean configuration unless the body is homogeneous. In calculating the components $F^m{}_{\alpha;\beta}$ one uses the affine connection $\boldsymbol{\Gamma}$ defined over the body manifold as before:

$$F^m{}_{\alpha;\beta} = \frac{\partial^2 x^m}{\partial X^\alpha \, \partial X^\beta} + \begin{Bmatrix} m \\ q \ r \end{Bmatrix} x^q{}_{,\alpha} \, x^r{}_{,\beta} - \Gamma_{\beta\alpha}{}^\sigma x^m{}_{,\sigma},$$

but now $\boldsymbol{\Gamma}$ cannot be derived from a metric tensor except in the case when the body is homogeneous, and, in particular, there is in general no co-ordinate system in which the components $\Gamma_{\beta\alpha}{}^\sigma$ vanish. When the dislocations are specified in the sense that $\boldsymbol{\Gamma}$ is a given function of position in the uniform reference, we thus obtain an explicit differential equation of motion. If $\boldsymbol{\Gamma}$ is symmetric, NOLL's equation reduces to the classic equation of finite elasticity. More generally, it provides a starting point from which, at long last, it will be possible to set up and solve problems in the sense that the word "problem" is used in other parts of mechanics but not as yet in dislocation theory, namely, to find the effect of applying forces or given states of motion to a body.

Once the core of the classical mechanics of materials had been straightened and annealed by these clear and embracing concepts and definitions, it became a mathematical science after the manner of geometry, with a breadth of application no competent critic denies. While the oldsters of "applied mechanics" shook their pates and tugged their smooth-shaven chins at the abstraction needed to contemplate a tensor-valued functional of tensor-histories of the deformation of a manifold, the practice of the subject poured out major theorems made possible by the simplicity of this very idea.

First, Messrs. COLEMAN & NOLL were able to calculate in full generality the steady viscometric flows of an arbitrary simple fluid. These flows are all of shearing type, within parallel plates or concentric cylinders or coaxial cones; they form the basis for the theory of viscometers. The geometry is such that an observer moving with a fluid particle can orient himself so as to see a constant deformation history behind him. Thus the fluid, while it may be capable of showing complicated effects of memory in general, in these particular flows is left little to remember. For this reason explicit and general solution of the viscometric problems, even for the most complicated fluids, is easy. Generalizing and unifying the great work of RIVLIN on particular fluids in 1947—1956, COLEMAN & NOLL proved in 1959 that the stress system in any simple fluid undergoing a viscometric flow is given explicitly by three scalar functions, whereby the entire viscometric program may be correlated and cross-checked. On the basis of this theory viscometry has become a coherent, closed science, which has been shown in the last year to explain satisfactorily the hitherto confused experiments on "non-Newtonian" fluids. Solutions of polyisobutylene in cetane, for example, show no discernible range of Newtonian behavior but obey perfectly the theory of COLEMAN & NOLL. A treatise on the subject, the first to relate theory and experiment adequately since the classic book of BRILLOUIN on Newtonian fluids, is now in press. The experimental side is handled by Mr. MARKOVITZ, who was a Visiting Professor at this university in 1958.

Second, Messrs. COLEMAN & NOLL formulated a general concept of fading memory in materials. Thus far, the response functional has been arbitrary, so that the theory is broad enough to include materials which allow a large influence to deformation in the distant past. COLEMAN & NOLL define the recollection of a history as the average value of its magnitude, weighted by an obliviator, which is a factor that diminishes the contribution of past events, the more so, ultimately, the longer ago they occurred. A material is said to have fading memory if its response functional is continuous at the rest history in the topology defined by recollection. Thus COLEMAN & NOLL's idea of fading memory calls for the stress to differ little from the equilibrium stress if the deformation has been small in the recent past. In effect, a material has fading memory if it shows stress relaxation. If instead of being merely continuous, the constitutive functional is differentiable, it may be approximated by a much simpler one, which COLEMAN & NOLL find as an explicit function of the present strain and stretching. In this way a rational position is found within NOLL's general ideas for all the classical theories of solids, fluids, and visco-elastic substances, and a method is provided for calculating higher approximations of similar type.

Because of the example of the Navier-Stokes fluid, it is widely believed that the presence of dissipative mechanisms such as viscosity and heat conduction instantly smoothes out any discontinuities that may form, rendering shock waves and acceleration fronts impossible. This is not so. The viscous fluid and its generalization, the Rivlin-Ericksen fluid, are exceptional materials in that they do not obey COLEMAN & NOLL's principle of fading memory. In this regard the infinitesimally visco-elastic material of BOLTZMANN is more typical. Within this special and approximate theory the existence of singular surfaces was discovered in 1949 by R. SIPS and 1953 by E. H. LEE & I. KANTER. Indeed, in 1930 LAMPARIELLO had remarked that dissipative mechanisms do not in general prevent the persistence of singular surfaces, adducing as his example a vibrating string with linear damping, where, as he proved, the speeds of propagation of acceleration waves are unaffected by the friction. A massive study, partly in press today and partly still in draught form, by Messrs. COLEMAN & GURTIN has shown that wave propagation is guaranteed by the principle of fading memory. Little need be known about the constitutive functional in order to calculate explicitly the speeds of various kinds of waves and their decay or growth in amplitude as they progress. The corresponding results for linearized visco-elasticity and for finite elastic strain appear as limit cases for materials having in a sense maximum or minimum (zero) internal friction. Thus expression of the principle of fading memory as a mathematical definition has led to the discovery of a general theory of waves in dissipative materials.

Finally, the theory of simple materials has suggested the creation of thermodynamics as a part of the rational science of deformation, as I will explain in the third lecture.

The audience will now find itself divided into two parts. The greater, being those persons not before now conversant with modern mechanics, will have been impressed, whether favorably or unfavorably — I hope not entirely the latter — with the extreme generality of what I have presented. The smaller, consisting of experts in the field, will complain that after barely alluding to multipolar media and thermodynamics, I have left unmentioned microstructure, diffusion, chemical reactions, electromagnetism, relativity, and perhaps still more that is their own daily food of study and research.

These lectures are intended for beginners. I have adopted a conservative position. There are times when generality is needed, and others when it is all too easy. In rational mechanics we are just seeing the change from one of these times to another. Formerly, the beginner was taught to crawl through the underbrush, never lifting his eyes to the trees; today he is often made to focus on the curvature of the universe, missing even the earth. It is now more important to consolidate the gains of the past fifteen years by clearer understanding of principle and by intensive mathematical analysis than to map out new terrain. In the following lectures I will take up some of the more general ideas the experts may have expected me to develop today, but again it is solid successes in specific problems, not pursuit of the cosmos, that will furnish my subjects.

II. Polar and oriented media

CAUCHY'S second law, that the stress tensor is symmetric:

$$T = T^T,$$

is a statement of the balance of moment of momentum in continuum mechanics. In fact, it expresses more, for it implies that all torques are moments of force. According to the usual stress principle, we imagine a surface which separates the material into two parts, and we regard the action of the material upon one side of it on that on the other side as being equipollent to the action of a field of stress vectors $t_{(n)}$ defined on the surface: By assertion, then, the torque per unit area exerted by these vectors about some point O is

$$p \times t_{(n)},$$

where p is the position vector from O to the point where $t_{(n)}$ acts. As was suggested by VOIGT in 1887, the action of material on

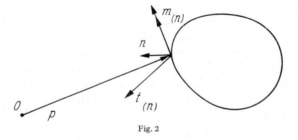

Fig. 2

material may give rise to couples as well as moments. If the couple per unit area is $m_{(n)}$, then the total torque per unit area is

$$p \times t_{(n)} + m_{(n)}.$$

Just as $t_{(n)}$ may be obtained from the unit normal n through a stress tensor T, $m_{(n)}$ may be obtained through a couple-stress tensor M:

$$t_{(n)} = Tn, \quad m_{(n)} = Mn.$$

The field equation expressing the principle of moment of momentum now takes a form resembling CAUCHY's first law:

$$\operatorname{div} \boldsymbol{T} + \varrho\,\boldsymbol{b} = \varrho\,\ddot{\boldsymbol{x}},$$

$$\operatorname{div} \boldsymbol{M} + \varrho\,\boldsymbol{l} = \lfloor \boldsymbol{T} \rfloor,$$

where

$$\lfloor \boldsymbol{T} \rfloor \equiv \tfrac{1}{2}\,(\boldsymbol{T} - \boldsymbol{T}^{T})$$

and where $\boldsymbol{l}$ is the density of body couple per unit mass. CAUCHY's
second law results when $\boldsymbol{M} = \boldsymbol{0}$ and $\boldsymbol{l} = \boldsymbol{0}$.

However, when we start to generalize, it is hard to know when
to stop. With torque goes moment of momentum, and we may
imagine that this, too, consists in two parts: To the moment of the
linear momentum $\boldsymbol{p} \times \varrho\,\dot{\boldsymbol{x}}$ is added a "spin angular momentum"
$\varrho\,\boldsymbol{s}$, independent of the motion of the point with which it is associated:

$$\boldsymbol{H} = \int_{\mathscr{B}} \varrho\,(\boldsymbol{p} \times \dot{\boldsymbol{x}} + \boldsymbol{s})\,dv.$$

Such an angular momentum is easily visualized by means of a
model in which the "particle" is not a point but an infinitesimally
small rigid body. The moment of momentum of a number of such
particles is not merely the sum of the moments of the linear
momenta of each, localized at the centers of mass, since the spins
of the small bodies contribute on an equal footing. When we wish to
represent by a continuum model the gross behavior of a material
the chemists tell us is composed of molecules having structure,
such as the favorite dumbbell, we may wish to take spin angular
momentum into account. If we do, we find that the field equation
is generalized again:

$$\operatorname{div} \boldsymbol{M} + \varrho\,\boldsymbol{l} = \varrho\,\dot{\boldsymbol{s}} + \lfloor \boldsymbol{T} \rfloor.$$

To consider this modified angular momentum at all, we have
implicitly presumed a more general body manifold. The configuration of a body is no longer a region of Euclidean space but
rather a region filled with triads of vectors. With each place in
the configuration are associated three vectors $\boldsymbol{d}_a$, the *directors*
of the medium, and accordingly the body manifold being mapped
into space is a set of elements, each of which is a point and three
vectors. If these vectors always occupy orthogonal configurations,

their angular velocity or spin can be defined, and the spin angular momentum is that which corresponds to this spin.

We have been accustomed to such a model by the classical theories of rods and shells. In order to account for the possibility of torsion in a one-dimensional theory, we introduce as a model a line, to each point of which a vector is attached:

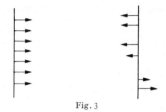

Fig. 3

Denoting by $\theta(s)$ the angle subtended upon a fixed direction by these vectors, we define the twist as $\theta'(s)$. Here the micro-structure assigned, by definition, to the rod results from trying to represent by a one-dimensional model a phenomenon we consider to be truly three-dimensional. DUHEM suggested in 1893 that the gross aspects of almost any kind of molecular constitution or other internal structure of a material in three dimensions could be represented by a continuum of points and directors — as we now call it, an *oriented medium*. Such a model, while not describing the individual motions of molecules at all, abstracts their contributions to the response and motion of the body as a whole. Of course the continuum theory can yield, in principle, less information about a material than does that of a correct, detailed molecular model. In the theory of rods, we agree in advance to be content with, at best, the right value for the twist, intentionally excluding all the other effects and properties a truly three-dimensional model might give us. No one expects the theory of elastic rods to tell him all that three-dimensional elasticity can yield, but, as every engineer knows, the advantage of the more refined model in principle becomes an illusion in a practical case where its equations are too complicated to yield specific results. Continuum and molecular models of three-dimensional bodies of any complexity stand in the same relation.

If the directors $\boldsymbol{d}_a$ are to represent structure, they need not be orthogonal or rigid, and their number may have to be greater than three. Mr. ERICKSEN and I considered and characterized the general strain of an oriented medium in 1958, but we stopped short of the dynamical equations and constitutive equations for special kinds of oriented materials, since we saw that the deformation of the directors ought to affect everything, but we shrank from excessive generality without a guide.

Mr. ERICKSEN in the following years proposed and studied a class of materials having a single director at each point. The theory was made as simple as seemed possible in view of its intended application to the mechanics of liquid crystals. ERICKSEN's analysis of special flows serves a model to show how in a rate theory phenomena of instability may result in what seems, superficially, a change of material properties, but it would be presumptuous of me to attempt to explain what we may all learn better directly from him. Let it here suffice to say that the aim of this special theory was to provide a specimen rather than to generalize or organize the field of oriented materials.

The best of several guides in forming the various non-linear theories organized and generalized by NOLL's theory of materials was classical finite elasticity. Such a guide for oriented materials was sketched by Mr. ERICKSEN in a lecture in 1962 which was not published. The idea was discovered independently by Mr. TOUPIN shortly thereafter, and he carried it through to a general and properly invariant theory of hyperelastic materials with microstructure.

A fully conservative situation can be described by an action principle, which has the advantage of making the theory accessible also to physicists. The specialist in mechanics, likely to prefer a more direct approach, may easily work it out from the results. TOUPIN supposes the action density L to depend upon place, directors, and the gradients and time rates of these quantities as well as, possibly, the time and the particle. If the motion is

$$x = \chi(X, t), \quad d_a = \chi_a(X, t),$$

set

$$F \equiv \nabla \chi, \quad F_a \equiv \nabla \chi_a,$$

where ∇ denotes the gradient with respect to X. F is the deformation gradient as usual, and F_a is the director gradient. Then the action associated with a body $\mathscr{B}$ in the interval $t_1 \leq t \leq t_2$ is given by

$$A = \int_{t_1}^{t_2} \int_{\mathscr{B}} L(\boldsymbol{x}, \boldsymbol{d}_a, \dot{\boldsymbol{x}}, \dot{\boldsymbol{d}}_a, \boldsymbol{F}, \boldsymbol{F}_a, \boldsymbol{X}, t)\, dv_R\, dt,$$

where the integration is carried out in the reference configuration, in which places are denoted by $\boldsymbol{X}$. The action principle follows the tradition of LAGRANGE and PIOLA in introducing indeterminate multipliers for each term that can arise in varying χ and χ_a independently:

$$0 = \delta A + \int_{t_1}^{t_2} \int_{\mathscr{B}} \varrho_R (\boldsymbol{b} \cdot \delta \chi + \boldsymbol{b}^a \cdot \delta \chi_a)\, dv_R\, dt +$$

$$+ \int_{t_1}^{t_2} \int_{\partial \mathscr{B}} \varrho_R (\boldsymbol{t}_R \cdot \delta \chi + \boldsymbol{t}_R^a \cdot \delta \chi_a)\, ds_R\, dt -$$

$$- \int_{\mathscr{B}} \varrho_R (\boldsymbol{m} \cdot \delta \chi + \boldsymbol{m}^a \cdot \delta \chi_a)\, dv_R \big|_{t_2}^{t_1},$$

where summation over a is understood. Suggesting the interpretation that will result, we name these multipliers:

$\boldsymbol{b}$ body macroforce density,
$\boldsymbol{b}^a$ body microforce density,
$\boldsymbol{t}_R$ macrostress vector in the reference configuration,
$\boldsymbol{t}_R^a$ microstress vector in the reference configuration,
$\boldsymbol{m}$ macromomentum per unit mass,
$\boldsymbol{m}^a$ micromomentum per unit mass.

In the formal Lagrangean way, we introduce symbols for the partial derivatives of L:

$$\varrho_R\, \boldsymbol{m}^* \equiv \frac{\partial L}{\partial \dot{\boldsymbol{x}}}, \qquad \boldsymbol{T}_R \equiv -\frac{\partial L}{\partial \boldsymbol{F}},$$

$$\varrho_R\, \boldsymbol{m}^{a*} = \frac{\partial L}{\partial \dot{\boldsymbol{d}}_a}, \qquad \boldsymbol{T}_R^a \equiv -\frac{\partial L}{\partial \boldsymbol{F}_a},$$

$$\varrho_R\, \boldsymbol{b}^{a*} \equiv -\frac{\partial L}{\partial \boldsymbol{d}_a}.$$

The Euler equations for the time limits yield the expected identity of the momenta with the corresponding gradients:

$$\boldsymbol{m}^* = \boldsymbol{m}, \quad \boldsymbol{m}^{a*} = \boldsymbol{m}^a,$$

when $t = t_1$ or t_2 and hence for all t. Using this fact, we express the Euler equations for the interior and the boundary as follows:

$$\text{Div } \boldsymbol{T}_R + \varrho_R\, \boldsymbol{b} = \varrho_R\, \dot{\boldsymbol{m}} - \frac{\partial L}{\partial \boldsymbol{x}}\,,$$

$$\text{Div } \boldsymbol{T}_R^a + \varrho_R(\boldsymbol{b}^a - \boldsymbol{b}^{a*}) = \varrho_R\, \dot{\boldsymbol{m}}^a\,,$$

and

$$\boldsymbol{T}_R\, \boldsymbol{n}_R = \boldsymbol{t}_R\,,$$

$$\boldsymbol{T}_R^a\, \boldsymbol{n}_R = \boldsymbol{t}_R^a\,.$$

As in all variational formulations, one needs to have a good idea by other means what the answer ought to be. Here we see forms similar to those of the equation of linear momentum and the associated boundary condition. To check, consider the classical theory of hyperelasticity, defined by the existence of a stored-energy function or stress potential $\sigma(\boldsymbol{F}, \boldsymbol{X})$ depending, at each particle, on the deformation gradient alone. To obtain this theory, we try a particular Lagrangean, the difference of the densities of kinetic and potential energy:

$$L = \varrho_R\left[\tfrac{1}{2}\, \dot{\boldsymbol{x}}^2 - \sigma(\boldsymbol{F},\, \boldsymbol{X})\right].$$

By definition, for this case, most of the quantities associated with the directors vanish:

$$\boldsymbol{m}^a = \boldsymbol{T}_R^a = \boldsymbol{b}^{a*} = 0\,,$$

and the density of linear momentum reduces to the velocity:

$$\boldsymbol{m} = \dot{\boldsymbol{x}}\,.$$

Still by definition, the quantity $\boldsymbol{T}_R$ becomes the elastic stress in the reference configuration, called the Piola stress tensor:

$$\boldsymbol{T}_R = \varrho_R\, \frac{\partial \sigma}{\partial \boldsymbol{F}}\,.$$

The first equations and boundary conditions of momentum type reduce to

$$\text{Div } \boldsymbol{T}_R + \varrho_R\, \boldsymbol{b} = \varrho_R\, \ddot{\boldsymbol{x}}\,,$$

$$\boldsymbol{T}_R\, \boldsymbol{n}_R = \boldsymbol{t}_R\,,$$

having the same form as CAUCHY's first law and boundary condition in the reference configuration; and the second set of momentum equations becomes

$$b^a = 0, \qquad t^a_R = 0,$$

asserting that the classical theory of elasticity can allow no body or surface microforces.

We are on the right track, but we do not yet have the theory of hyperelasticity, for we lack two of its major principles: The Cauchy stress tensor T is symmetric, and the stored-energy function depends on the deformation gradient F only through the strain tensor $C = F^T F$. It is known that these two principles are equivalent to each other and to the invariance of σ under constant rigid displacements. This *Euclid invariance* is a special case of the principle of material frame-indifference; for elasticity, the full strength of that principle is not needed. In proposing and exploring a theory similar to TOUPIN's but restricted to the case of a rigid orthonormal director frame, a theory they called "Euclidean action", E. & F. COSSERAT in 1907 had laid down the principle that L be Euclid invariant. Their profound work attracted little attention in its own day and was soon forgotten. Much of the COSSERATS' effort concerned rods and shells, but the engineers of the period may not have had enough interest in an exact theory to follow so many pages of long calculation in general terms. The final end the COSSERATS seem to have had in mind for their three-dimensional model was an aether and radiation theory, replacing and uniting the Fresnel-Cauchy elastic theory and the MacCullagh-Maxwell theory. Although the COSSERATS' book was first published as an appendix to an influential treatise on physics as a whole, the physicists of the day may not have appreciated the careful statement of hypotheses and rigorous mathematics that extended the work past 200 pages. The COSSERATS' masterpiece stands as a tower in the field. It was noticed, cited, but not read in detail by the half dozen people responsible for the resurgence of rational mechanics fifteen to twenty years ago. Had we mastered the analysis of the COSSERATS then, not only would time and work of rediscovery have been spared, but also a paragon of method would have lain in our hands.

In his more general theory of oriented hyperelastic materials, Toupin lays down the principle of Euclid invariance as fundamental. In order that the action be invariant under constant rigid displacement, it must be invariant under infinitesimal rigid displacements in particular, and these may be decomposed into three kinds:

1. $\delta\boldsymbol{\chi}=$const., $\delta\boldsymbol{\chi}_a=\boldsymbol{0}$ (translation with fixed directors).

2. $\delta\boldsymbol{\chi}=\boldsymbol{W}\,\boldsymbol{p}$, $\delta\boldsymbol{\chi}_a=\boldsymbol{W}\,\boldsymbol{d}_a$, where $\boldsymbol{W}=-\boldsymbol{W}^T$ (rotation of body and directors together).

3. $\delta\boldsymbol{\chi}=\dot{\boldsymbol{x}}$, $\delta\boldsymbol{\chi}_a=\dot{\boldsymbol{d}}_a$ (translation along the velocity).

It turns out that necessary and sufficient conditions for invariance of the action under these three special transformations, and hence under all constant rigid motions, are

$$1. \quad \frac{\partial L}{\partial \boldsymbol{x}}=\boldsymbol{0},$$

$$2. \quad \boldsymbol{K}=\boldsymbol{K}^T,$$

$$3. \quad \frac{\partial L}{\partial t}=0,$$

where

$$\varrho_R \boldsymbol{K}=\boldsymbol{p}\otimes\frac{\partial L}{\partial \boldsymbol{x}}+\boldsymbol{d}_a\otimes\frac{\partial L}{\partial \boldsymbol{d}_a}+\dot{\boldsymbol{x}}\otimes\frac{\partial L}{\partial \dot{\boldsymbol{x}}}+$$
$$+\boldsymbol{d}_a\otimes\frac{\partial L}{\partial \dot{\boldsymbol{d}}_a}+\boldsymbol{F}\left(\frac{\partial L}{\partial \boldsymbol{F}}\right)^T+\boldsymbol{F}_a\left(\frac{\partial L}{\partial \boldsymbol{F}_a}\right)^T.$$

Condition 2, asserting that $\boldsymbol{K}$ is a symmetric tensor, corresponds to the classical condition that the Cauchy stress tensor be symmetric, reducing to it when L has the form appropriate to classical elasticity. Consequently, we expect that Condition 2 should express the principle of moment of momentum in general, and that Conditions 1 and 3 should express the principles of linear momentum and energy. This expectation is correct, and it guides us in definitions of what the densities of moment of momentum and energy should be. Namely,

$$\boldsymbol{A}\equiv\{\boldsymbol{p}\otimes\boldsymbol{m}+\boldsymbol{d}_a\otimes\boldsymbol{m}^a\}=-\boldsymbol{A}^T,$$
$$E\equiv\boldsymbol{m}\cdot\dot{\boldsymbol{x}}+\boldsymbol{m}^a\cdot\dot{\boldsymbol{d}}_a-\frac{L}{\varrho_R}.$$

With these definitions, it turns out that the original action principle, now rendered Euclid invariant, holds if and only if

$$\int_{\mathcal{B}} \varrho_R \, \boldsymbol{m} \, dv_R \big|_{t_1}^{t_2} = \int_{t_1}^{t_2} \Big[\int_{\mathcal{B}} \varrho_R \, \boldsymbol{b} \, dv_R + \int_{\partial\mathcal{B}} \boldsymbol{t}_R \, ds_R \Big] \, dt,$$

$$\int_{\mathcal{B}} \varrho_R \, \boldsymbol{A} \, dv_R \big|_{t_1}^{t_2} = \int_{t_1}^{t_2} \Big[\int_{\mathcal{B}} \varrho_R \, \{\boldsymbol{p} \otimes \boldsymbol{b} + \boldsymbol{d}_a \otimes \boldsymbol{b}^a\} \, dv_R +$$
$$+ \int_{\partial\mathcal{B}} \{\boldsymbol{p} \otimes \boldsymbol{t}_R + \boldsymbol{d}_a \otimes \boldsymbol{t}_R^a\} \, ds_R \Big] \, dt,$$

$$\int_{\mathcal{B}} \varrho_R \, E \, dv_R \big|_{t_1}^{t_2} = \int_{t_1}^{t_2} \Big[\int_{\mathcal{B}} \varrho_R (\boldsymbol{b} \cdot \dot{\boldsymbol{x}} + \boldsymbol{b}^a \cdot \dot{\boldsymbol{d}}_a) \, dv_R +$$
$$+ \int_{\partial\mathcal{B}} (\boldsymbol{t}_R \cdot \dot{\boldsymbol{x}} + \boldsymbol{t}_R^a \cdot \dot{\boldsymbol{d}}_a) \, ds_R \Big] \, dt.$$

These equations assert that the increase in linear momentum, moment of momentum, and energy in the time between t_1 and t_2 is exactly equal to the time-integrated applied force, moment of applied force, and rate of working of applied force, as we agree to picture the microforces as acting at the ends of the directors. The typical dynamic element of the body in the reference configuration may be visualized as shown:

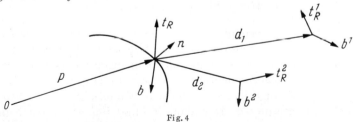

Fig. 4

where only two directors are drawn. The result established may be called the *Cosserat-Toupin theorem: in an oriented hyperelastic material subject to the action principle postulated, the action density is Euclid invariant if and only if linear momentum, moment of momentum, and energy are balanced.*

This is only the beginning of Toupin's theory, since all that has been accomplished so far is to set up the equations in terms of a reference configuration. It remains to transfer all equations to the deformed configuration of the body, to interpret the results, and to solve important specific problems. There is no great difficulty, but time does not permit me to present the details. Most important is the occurrence of couple stresses as a necessary consequence of what has gone before. Indeed, if we define the third-order tensor of *hyperstress* $\boldsymbol{H}$:

$$H_{km}{}^{p} \equiv -\frac{\varrho}{\varrho_{\mathrm{R}}}\, d_{ak}\, \frac{\partial L}{\partial d_{a,\alpha}^{m}}\, x^{p}{}_{,\alpha},$$

the second field equation can be shown to take the following form in the deformed configuration of the body:

$$\mathrm{div}\ \boldsymbol{H} + \boldsymbol{T}^{T} + \varrho\boldsymbol{L} = \varrho\,(\dot{\boldsymbol{S}} - \boldsymbol{K} + \dot{\boldsymbol{x}}\otimes\boldsymbol{m}),$$

where

$$\boldsymbol{S} \equiv \boldsymbol{d}_{a}\otimes\boldsymbol{m}^{a}, \quad \boldsymbol{L} \equiv \boldsymbol{d}_{a}\otimes\boldsymbol{b}^{a}.$$

As we have seen, $\{\boldsymbol{S}\}$ appears as the second term in $\boldsymbol{A}$, being the directors' contribution to the total moment of momentum. If we define the *couple-stress tensor* $\boldsymbol{M}$ as the skew part of $\boldsymbol{H}$:

$$M^{kpq} \equiv H^{[kp]q},$$

then the skew part of the second field equation becomes

$$\mathrm{div}\ \boldsymbol{M} + \varrho\{\boldsymbol{L}\} = \varrho\{\dot{\boldsymbol{S}} + \dot{\boldsymbol{x}}\otimes\boldsymbol{m}\} + \{\boldsymbol{T}\}.$$

This equation we recognize as the final generalization of CAUCHY's second law, since it contains the contributions to the couple stress arising from change of spin angular momentum and from possible failure of the linear momentum $\boldsymbol{m}$ to be parallel to the velocity $\dot{\boldsymbol{x}}$.

The theory of oriented hyperelastic materials is now clearly stated and interpreted. It remains to find important particular solutions and develop their physical interpretations, with a view to explaining observed phenomena in solids excluded from the classical theory of elasticity, such as the dispersion of plane waves in an infinite medium. As often happens, some of these results had already been obtained on a less secure basis, and more are now being found. Mr. MINDLIN and his students have calculated a number of special solutions for infinitesimal strain and have interpreted them, particularly in the context of vibrations of various anisotropic materials. I turn instead to a different theory aimed at the same class of phenomena, namely, the theory of elastic materials of second grade, due also to TOUPIN.

In the theory of simple materials, presented in Lecture I, the stress is assumed to be determined by the history of the deformation gradient. In oriented materials, the body manifold itself is generalized by the addition of directors at the points, but the point deformation still affects the stress only through its gradient. A different generalization is equally plausible. Leaving the body manifold as it is, a set of points only, we may allow the stress to depend in a more delicate way upon the deformation. NOLL'S general theory of materials allows all properties of the deformation in a neighborhood of a point to influence the stress there, but from the outset the stress tensor is assumed symmetric. Mr. TOUPIN has constructed a theory of elastic materials that may have unsymmetric stress tensors — polar-elastic materials of second grade.

In this theory the stored energy is allowed to depend upon F and ∇F:

$$E(\mathscr{B}) = \int_{\mathscr{B}} \varrho_R \sigma(F, \nabla F, X)\, dv_R,$$

and the mechanical axiom is the principle of virtual work:

$$\delta E = \int_{\mathscr{B}} \varrho_R f \cdot \delta\chi\, dv_R + \int_{\partial\mathscr{B}} (t_R \cdot \delta\chi + h_R \cdot D\delta\chi)\, ds_R,$$

where h_R, like t_R, is an undetermined multiplier, and $D\delta\chi$ is the derivative of $\delta\chi$ in the direction normal to the boundary. The manipulations leading to equations of motion and boundary conditions are somewhat more elaborate than for the theory of oriented hyperelastic materials, and I omit them. Again a theory with couple stress results, but this time the boundary condition to be satisfied by the ordinary stress is influenced by the curvature of the surface on which it is applied, so that an effect of "puckering" of a boundary free of ordinary traction becomes possible.

To close the lecture, I remark upon an important consequence of comparing these two immediate generalizations of the classical theory. While they rest upon different basis concepts, their results are similar in form. Which of them is "correct"? Commonly such a question is decided by an appeal to experiment. If the predictions of a theory are confirmed by measured data, physical scientists usually hail the correctness of the hypotheses. In the present case, we are to decide whether a given specimen "really is" endowed with the microstructure of an oriented medium, or "really is" a point manifold in which the second gradient of the deformation contributes to the energy stored, or, possibly, neither of these. The question is undecidable. Not only are the two theories similar; in a major special case they become *identical*. In this case, experiment cannot agree with one without agreeing with the other. Both models are idealizations of a structure that is, doubtless, too complicated to be described perfectly by any theory. Which is to be used, in the cases when use of either, and hence both, is justified, remains a matter of convenience, or, perhaps, taste. Recall the similar case of optical waves in empty space, where MacCullagh's theory of the quasi-elastic aether and Maxwell's electromagnetic theory, although starting from different ground concepts, lead to identical predictions. It is impossible to decide between those theories on the basis of experiment. The universal preference given to Maxwell's theory cannot be explained except on grounds outside the theory itself and the experiments it explains. At worst, these grounds lie in the distaste physicists have been taught to experience when they hear the word "aether"; at best, they rest on the fertility of Maxwell's views as seeds for more general theories. In the case of polar-elastic materials, such grounds are wanting. Since both theories were developed by the same man, we cannot even attribute their diverse motivations to socio-economic injustice, wrangling among colleagues, or an unhappy love affair. That two theories of such sophistication for a common range of phenomena have been developed in a few pages and simultaneously indicates the power and flexibility of the new methods in rational mechanics.

III. Thermodynamics of visco-elasticity

Thermostatics, which even now is usually called thermodynamics, has an unfortunate history and a cancerous tradition. It arose in a chaos of metaphysical and indeed irrational controversy, the traces of which drip their poison even today. As compared with the older science of mechanics and the younger science of electromagnetism, its mathematical structure is meager. Though claims for its breadth of application are often extravagant, the examples from which its principles usually are inferred are most special, and extensive mathematical developments based on fundamental equations, such as typify mechanics and electromagnetism, are wanting. The logical standards acceptable in thermostatics fail to meet the criteria of other exact sciences; in books and papers concerning it the proportion of words if not prayers to equations is high — to proved theorems, almost infinite. There is nothing that can be said by mathematical symbols and relations which cannot also be said by words. The converse, however, is false. Much that can be and is said by words cannot successfully be put into equations, because it is nonsense. When a physical writer expresses an assertion in words only, he is refusing to stand up to the test. The early studies of thermodynamics abound in nonsensical wording, some of which the wars among the creators of the subject served to clear away, but they left much policing for later generations to do. As the area receded from the frontier of physical thought, however, it fell into the hands of text compilers, always eager to water and "explain" what is generally accepted without re-creating or even criticising it, and by their care some of the confusion such writers as Boltzmann succeeded in eliminating has been reintroduced for the help of students. The obscurity of all that concerns heat and temperature is witnessed by the many attempts, made by scientists of several brands and continuing today in greater number, to set the house of thermostatics in order.

The phenomena of gain and loss of heat and work cry for a true thermodynamics, a science of energy coupled with motion, in which dissipation is the rule, not the exception, and the effects of dissipation are calculated explicitly, not merely dismissed by attaching a plus or minus sign.

In the past year, a true thermodynamics of deformable media has been attained. The first step, as earlier for pure mechanics, was to renounce pursuit of the universe (a renunciation particularly difficult for thermodynamicists) and to solve a specific, concrete problem, but solve it well. The problem selected by COLEMAN & NOLL in 1963, when both were visiting professors at this university, was to find a fully thermodynamic theory of perfectly elastic materials endowed with linear viscosity. Here the equations desired were all known in advance. The problem was to derive them, within the general framework of mechanics, from appropriate general principles. The laws of mechanics, assumed to hold from the outset, are

1. Conservation of Mass.
2. Balance of Linear Momentum.
3. Balance of Moment of Momentum.
4. Balance of Energy.

In particular, we have a mass density ϱ and a stress tensor $\boldsymbol{T}$, in terms of which the principles may be expressed in their classical forms:

$$\dot{\varrho} + \varrho \operatorname{div} \dot{\boldsymbol{x}} = 0,$$

$$\operatorname{div} \boldsymbol{T} + \varrho \boldsymbol{b} = \varrho \ddot{\boldsymbol{x}},$$

$$\boldsymbol{T} = \boldsymbol{T}^{\scriptscriptstyle T},$$

$$\operatorname{tr}(\boldsymbol{T}\,\boldsymbol{D}) + \operatorname{div} \boldsymbol{h} + \varrho q = \varrho \dot{\varepsilon},$$

where the notations are as follows:

ε specific internal energy,

$\boldsymbol{D}$ stretching tensor, $\frac{1}{2}(\boldsymbol{L} + \boldsymbol{L}^{\scriptscriptstyle T})$, where $\boldsymbol{L} = \operatorname{grad} \dot{\boldsymbol{x}}$,

q specific heat absorption.

The heat absorption q may be interpreted as a supply of energy in the interior by radiation and in other ways. It is important for the theory as being a quantity analogous to the body force $\boldsymbol{b}$. The possibility of assigning these two quantities arbitrarily, at least in theory, shows that the principles of momentum, moment of momentum, and energy in themselves *impose no restriction* on the variety of fields $\boldsymbol{F}$, $\dot{\boldsymbol{x}}$, $\boldsymbol{L}$ (and hence $\dot{\boldsymbol{F}}$), $\boldsymbol{T}$, ε, and $\boldsymbol{h}$.

Two additional variables are introduced to describe the effects of heat:

i) The *temperature* $\theta(\boldsymbol{x}, t)$, a time-dependent scalar field having positive values at every place $\boldsymbol{x}$ and time t.

ii) The *entropy* $H(\mathcal{B}, t)$ of the body $\mathcal{B}$ at the time t, a time-dependent additive set function of bodies.

We shall assume that H is an absolutely continuous function of mass, so that it has a density $\eta(\boldsymbol{x}, t)$, the specific entropy (per unit mass) at the place $\boldsymbol{x}$ and the time t.

In thermodynamic writings going back a hundred years inequalities restricting the growth of entropy were proposed. Mr. TOUPIN & I in 1960 laid down and called "the Clausius-Duhem inequality" the following:

$$\dot{H} \geqq \int_{\partial \mathcal{B}} \frac{\boldsymbol{h} \cdot \boldsymbol{n}}{\theta} \, ds + \int_{\mathcal{B}} \varrho \, \frac{q}{\theta} \, dv.$$

In this form of the *principle of entropy growth*, more general than any proposed before, sources of energy *of every kind* are assumed to contribute to the entropy *at least* as much as their ratios to the temperature. An equivalent local inequality is

$$\dot{\eta} \geqq \frac{1}{\varrho \theta} \operatorname{div} \boldsymbol{h} + \frac{q}{\theta} - \frac{1}{\varrho \theta^2} \, \boldsymbol{h} \cdot \operatorname{grad} \theta.$$

I turn aside to remark that the terms "first and second laws of thermodynamics" are not used, for the reason that these words mean different things to different people. For example, many books present the "first law" as a statement, in effect, that a "caloric equation of state" exists. Such an assumption, made also by the school of linearly irreversible thermodynamicists, I regard as special and hence expressly avoid among the basic principles. Indeed, one of the objectives of the general theory, which I shall mention at the end, will be to prove that a caloric equation of state in the usual sense exists only as an approximation. A result of this kind is suggested by a counterpart in a particular statistical theory: PRIGOGINE found that in the kinetic theory of monatomic gases the formal expansion of CHAPMAN & ENSKOG delivers a caloric equation of state at the stages of orders 0, 1, and 2, but not a at higher stage.

Some persons might call the Clausius-Duhem inequality "the second law of thermodynamics". That would be a fine term if only everyone else would agree to use it in the same sense and stop talking about universes running backward and forward reversibly or irreversibly, heat reservoirs, ideal steam engines, boilers, condensers, isolated systems, adiabatic and quasi-static processes, *etc.* For rational thermodynamics it is necessary above all else to puff away the blather and set down some concrete relations among mathematical quantities once and for all, as in other parts of mathematical physics. The equation of balance of energy and the Clausius-Duhem inequality for growth of entropy serve this purpose. No other general principles are required. As we shall see, however, the two laws stand at different stations. The principle of energy has already been listed alongside the principles of mass, momentum, and linear momentum. In COLEMAN & NOLL's approach, the principle of entropy growth is of a different kind.

Returning to the theory itself, I remark that none of the equations written down so far were new when COLEMAN & NOLL began their work. Their departure, to some degree anticipated in a letter sent me a little earlier by Mr. A. E. GREEN, was to demand that the Clausius-Duhem inequality be an *identity* for constitutive equations rather than a *condition* to be used in determining solutions to particular cases defined by prescribing particular forces and energies. COLEMAN & NOLL postulated that all thermo-mechanical constitutive equations must be such as to satisfy identically *two* requirements:

I. The principle of material frame-indifference.

II. The principle of entropy growth.

Since the body force b and the heat absorption q here are regarded as assignable arbitrarily in principle, while in any particular application they will be specified uniquely as a part of the definition of the problem, we may express COLEMAN & NOLL's basic assumption in the following alternative form, using the classical concept of "virtual" changes: *Every constitutive equation must be such as to satisfy both the principle of material frame-indifference and the Clausius-Duhem inequality for all virtual histories of deformation and temperature.*

COLEMAN & NOLL showed that the principle of entropy growth *severely restricts the possible constitutive equations.* They set up for illustrative study a special, simple class of materials, namely, those defined by constitutive equations of the form

$$T = \mathfrak{g}(F, \eta) + V(F, \eta)[L],$$

$$h = \hat{h}(F, \eta, \operatorname{grad} \theta),$$

where $V(F, \eta)[L]$ is a linear tensor function of L, depending upon F and η as parameters. The second equation generalizes FOURIER'S law of heat conduction by allowing the deformation to influence the thermal response of the material. If $V \equiv 0$ and $\eta = $ constant, the first equation is appropriate to the classical theory of finite elastic strain, while if $\mathfrak{g}$ and V depend upon F only through $\det F$, it defines the classical theory of linearly viscous fluids. For this latter case, V must equal an isotropic function of D, and the Clausius-Duhem inequality, as is well known, is equivalent to the statement that the shear and bulk viscosities cannot be negative. For finite elasticity, the energy *equation* rather than the entropy inequality was known to require that $\mathfrak{g}$ be derivable from a potential, called the stored-energy function, or, in other terms, that the elastic material be hyperelastic. COLEMAN & NOLL'S problem was to interpolate between these two limit cases, using a uniform method resting upon both of the thermodynamic principles. As in the limiting cases, it is assumed in this special theory that caloric equations of state exist:

$$\varepsilon = \hat{\varepsilon}(F, \eta),$$

$$\theta = \hat{\theta}(F, \eta).$$

Once the problem is seen clearly, its solution is not hard. Substituting the equation of energy into the inequality for the entropy field yields

$$\varrho(\theta\dot{\eta} - \dot{\varepsilon}) + \operatorname{tr}(T D) + \frac{1}{\theta} h \cdot \operatorname{grad} \theta \geqq 0.$$

The functions $\mathfrak{g}$, V, $\hat{h}$, $\hat{\varepsilon}$, $\hat{\theta}$ occurring in the constitutive equations must be such as to ensure that this inequality is satisfied in every thermo-kinematic change.

Consider first an isothermal process: $\operatorname{grad} \theta = 0$. Substituting and carrying out the differentiations, we find that

$$\operatorname{tr}\left\{\left[\mathfrak{g}(F, \eta) - \varrho F\left(\frac{\partial \hat{\varepsilon}(F, \eta)^T}{\partial F}\right)\right] L\right\} + \left(\hat{\theta}(F, \eta) - \frac{\partial \hat{\varepsilon}(F, \eta)}{\partial \eta}\right)\dot{\eta} +$$

$$+ \operatorname{tr}\{L \, V(F, \eta)[L]\} \geqq 0,$$

identically in the variables $F(t)$ and $\eta(t)$. First suppose $L = 0$. Then

$$\left(\hat{\theta}(F, \eta) - \frac{\partial\hat{\varepsilon}(F, \eta)}{\partial\eta}\right)\dot{\eta} \geqq 0.$$

Since $\dot{\eta}$ is arbitrary, it follows that the quantity in parentheses vanishes. The result is the classical formula for the temperature:

$$\theta = \hat{\theta}(F, \eta) = \frac{\partial\hat{\varepsilon}(F, \eta)}{\partial\eta},$$

and the corresponding term drops out of the inequality. Now we remark that if the inequality is to hold for all F and L, it must hold identically in α if we replace L by αL:

$$\alpha\,\mathrm{tr}\left\{\left[\mathfrak{g}(F, \eta) - \varrho F\left(\frac{\partial\hat{\varepsilon}(F, \eta)}{\partial F}\right)^T\right]L\right\} + \alpha^2\,\mathrm{tr}\{LV(F, \eta)[L]\} \geqq 0,$$

since $V(F, \eta)[L]$ is linear in L. The quadratic polynomial $\alpha A + \alpha^2 B$ is positive semi-definite if and only if $A = 0$ and $B \geqq 0$. Hence, first,

$$\mathrm{tr}\left\{\left[\mathfrak{g}(F, \eta) - \varrho F\left(\frac{\partial\hat{\varepsilon}(F, \eta)}{\partial F}\right)^T\right]L\right\} = 0.$$

By hypothesis, the value of $\mathfrak{g}$ is a symmetric tensor. From the requirement that the relation $\varepsilon = \hat{\varepsilon}(F, \eta)$ satisfy the principle of material frame-indifference, it is easy to show that $\hat{\varepsilon}(F, \eta) = \bar{\varepsilon}(F^T F, \eta)$ and hence that

$$F\left(\frac{\partial\hat{\varepsilon}(F, \eta)}{\partial F}\right)^T$$

is a symmetric tensor. Thus the last equation is of the form

$$\mathrm{tr}(Y\,D) = 0,$$

for arbitrary symmetric tensors D. The symmetric tensor Y cannot satisfy this condition unless it vanishes. Hence

$$\mathfrak{g}(F, \eta) = \varrho F\left(\frac{\partial\hat{\varepsilon}(F, \eta)}{\partial F}\right)^T.$$

This equation has the same form as the stress-strain relation of hyperelasticity. It asserts that $\mathfrak{g}(F, \eta)$, which is the portion of the stress corresponding to the static deformation, is derivable from a stored-energy function, and that function is precisely the internal energy function $\hat{\varepsilon}$.

The remaining inequality is

$$\mathrm{tr}\{\boldsymbol{L}\,\mathsf{V}(\boldsymbol{F},\eta)\,[\boldsymbol{L}]\}\geqq 0,$$

asserting that the rate of working of the viscous part of the stress is non-negative. Finally, if $\boldsymbol{L}=\boldsymbol{0}$ the original inequality reduces

$$\boldsymbol{h}\cdot\mathrm{grad}\;\theta\geqq 0,$$

stating that heat never flows against a temperature gradient.

We have shown that the temperature and stress-strain equations and the dissipation and thermal conduction inequalities are necessary consequences of the assumed principle of growth of entropy. Conversely, they are sufficient for it to hold.

To complete the formulation of the theory, we need to be sure that the constitutive equations satisfy the principle of material frame-indifference. To this end, as was remarked already, the internal-energy function must be of the form

$$\varepsilon=\hat{\varepsilon}\,(\boldsymbol{F},\eta)=\bar{\varepsilon}\,(\boldsymbol{F}^{T}\boldsymbol{F},\eta)\,.$$

Similar restrictions are easily derived for the functions giving the heat flux and the dissipative or viscous part of the stress:

$$\boldsymbol{h}=\boldsymbol{F}\,\bar{\boldsymbol{h}}\,(\boldsymbol{F}^{T}\boldsymbol{F},\eta,\boldsymbol{F}^{T}\,\mathrm{grad}\;\theta)\,,$$

$$\mathsf{V}(\boldsymbol{F},\eta)\,[\boldsymbol{L}]=\boldsymbol{F}\,\{\mathsf{V}(\boldsymbol{F}^{T}\boldsymbol{F},\eta)\,[\boldsymbol{F}^{T}\boldsymbol{D}\boldsymbol{F}]\}\,\boldsymbol{F}^{T}\,.$$

With $\bar{\varepsilon}$, $\bar{\boldsymbol{h}}$, and $\bar{\mathsf{V}}$ of this kind, provided the last two are such as to satisfy the respective dissipation inequalities:

$$\boldsymbol{h}\cdot\mathrm{grad}\;\theta\geqq 0,\qquad\mathrm{tr}\{\boldsymbol{L}\,\mathsf{V}(\boldsymbol{F},\eta)\,[\boldsymbol{L}]\}\geqq 0,$$

and with θ and $\boldsymbol{T}$ given by the temperature and stress relations:

$$\theta=\frac{\partial\hat{\varepsilon}}{\partial\eta}\,,\qquad\boldsymbol{T}=\varrho\,\boldsymbol{F}\left(\frac{\partial\hat{\varepsilon}}{\partial\boldsymbol{F}}\right)^{T}+\mathsf{V}(\boldsymbol{F},\eta)\,[\boldsymbol{L}]\,,$$

we have found the *general solution* of the equations and inequalities laid down as basic, namely, the principles of material frame-indifference and of entropy growth.

The theory of elastic materials with linear viscosity as just developed affords an example of how to reduce the apparent generality of an assumption to a special one by demanding consistency with overriding general principles. With the method now illustrated by application to a clear and familiar special case, let us look again at our starting point, which was the following constitutive equations:

$$T = \mathfrak{g}(F, \eta) + V(F, \eta)\,[L],$$

$$h = \hat{h}(F, \eta, \operatorname{grad} \theta),$$

$$\varepsilon = \hat{\varepsilon}(F, \eta),$$

$$\theta = \hat{\theta}(F, \eta).$$

These equations are unsymmetric in that L is allowed to affect only T, not h or ε or θ, while grad θ affects only h, none of the other variables. This separation of temperature gradient as the "cause" of heat flux but not of stress, and velocity gradient as the "cause" of stress but not of heat flux, is unnatural and unjustified by physical principle. Resulting only from the gradual discovery of individual phenomena, it reflects old opinions that break physics up into compartments. Theorists should not propose constitutive equations which artificially divert theories into disjoint channels.

The same view results also from statistical models of continuous matter. There, stress and heat flux appear as gross or mean expressions of a purely mechanical process. The separation of stress as arising from deformation and flux of energy from changes of temperature emerges, not without reason, as a first approximation, but on a finer scale, it is illusory.

Thus, whether we appeal ultimately to a continuum or a molecular model, we are led to impose the *principle of equipresence: A quantity present as an independent variable in one constitutive equation is so present in all, to the extent that its appearance is not forbidden by the general laws of physics or rules of invariance.*

Let is not be thought that this principle would invalidate the classical separate theories in the cases for which they are intended, or that no separation of effects remains possible. Quite the reverse: The various principles of invariance, stated above, when brought to bear upon a general constitutive equation have the effect of restricting the manner in which a particular variable, such as the spin tensor or the temperature gradient, may occur. The classical separations may always be expected, in one form or another, for small changes — not as assumptions, but as proven consequences of invariance requirements. The principle of equipresence states, in effect, that no division of phenomena is to be laid down by constitutive equations. It may be regarded as a natural extension of OCKHAM's razor as restated by NEWTON: "We are to admit no more causes of natural things than such as are both true and sufficient to explain their appearances, for nature is simple and affects not the pomp of superfluous causes." This more general approach has the added value of showing in what way the classical separations fail to hold when interactions actually occur.

To satisfy the principle of equipresence, we replace the thermo-elastic constitutive equations by the following more general ones:

$$\theta = \breve{\theta}\,(\varepsilon,\,\mathrm{grad}\,\theta,\,\boldsymbol{F},\,\dot{\boldsymbol{F}})\,,$$

$$\eta = \breve{\eta}\,(\varepsilon,\,\mathrm{grad}\,\theta,\,\boldsymbol{F},\,\dot{\boldsymbol{F}})\,,$$

$$\boldsymbol{T} = \breve{\mathfrak{g}}\,(\varepsilon,\,\mathrm{grad}\,\theta,\,\boldsymbol{F},\,\dot{\boldsymbol{F}})\,,$$

$$\boldsymbol{h} = \breve{\boldsymbol{h}}\,(\varepsilon,\,\mathrm{grad}\,\theta,\,\boldsymbol{F},\,\dot{\boldsymbol{F}})\,,$$

COLEMAN & MIZEL have considered this theory in a recent work. Analysis of the consequences of the principle of growth of entropy shows easily that the constitutive equations for θ and η reduce to the classical form of "caloric equations of state":

$$\eta = \breve{\eta}\,(\varepsilon,\,\boldsymbol{F})\,,\qquad \theta = \breve{\theta}\,(\varepsilon,\,\boldsymbol{F}) = \frac{1}{\dfrac{\partial \breve{\eta}\,(\varepsilon,\,\boldsymbol{F})}{\partial \varepsilon}}\,,$$

or, equivalently,

$$\varepsilon = \bar{\varepsilon}\,(\eta,\,\boldsymbol{C})\,,\qquad \theta = \bar{\theta}\,(\eta,\,\boldsymbol{C}) = \frac{\partial \bar{\varepsilon}\,(\eta,\,\boldsymbol{C})}{\partial \eta}\,,$$

where $\boldsymbol{C} = \boldsymbol{F}^{T}\boldsymbol{F}$. Thus the greater generality of response allowed for the temperature, entropy, and energy is illusory: Within the framework here set up, these quantities must obey the classical thermodynamic relations. Moreover, as in the simpler theory considered before, the static part of the stress is determined from $\bar{\varepsilon}$:

$$\boldsymbol{T} = \boldsymbol{T}^{0} + \boldsymbol{T}^{\mathrm{D}}\,,$$

$$\boldsymbol{F}^{T}\boldsymbol{T}^{0}\boldsymbol{F} = 2\varrho\,\boldsymbol{C}\,\frac{\partial \bar{\varepsilon}\,(\eta,\,\boldsymbol{C})}{\partial \boldsymbol{C}}\,\boldsymbol{C}\,.$$

However, the dissipative stress and the heat flux are no longer determined as effects arising from distinct causes. We find only the restrictions imposed by material frame-indifference, namely

$$\boldsymbol{F}^T \, \boldsymbol{T}^{\mathrm{D}} \, \boldsymbol{F} = \mathfrak{g}^{\mathrm{D}} (\eta, \, \boldsymbol{F}^T \, \mathrm{grad} \; \theta, \, \boldsymbol{C}, \, \boldsymbol{F}^T \boldsymbol{D} \boldsymbol{F}) \, ,$$

$$\boldsymbol{h} = \boldsymbol{F} \, \bar{\boldsymbol{h}} (\eta, \, \boldsymbol{F}^T \, \mathrm{grad} \; \theta, \, \boldsymbol{C}, \, \boldsymbol{F}^T \boldsymbol{D} \boldsymbol{F}) \, ,$$

and the dissipation inequality is

$$\mathrm{tr} \, (\boldsymbol{T}^{\mathrm{D}} \, \boldsymbol{D}) + \frac{1}{\theta} \, \boldsymbol{h} \cdot \mathrm{grad} \; \theta \geqq 0 \, .$$

The classical separation of effects follows only for the static portion of the stress, $\boldsymbol{T}^0$, not for the dissipative part $\boldsymbol{T}^{\mathrm{D}}$, and not for the heat flux, and the dissipation inequality does not generally split into two.

We have not exhausted the possibilities, however. No symmetry has been ascribed to the material, and no limitation has been placed upon the processes to which it is subject. COLEMAN & MIZEL have considered in greater detail the case of a fluid, defined as a material for which the isotropy groups of all four constitutive equations are the unimodular group, and have shown that then

$$\varepsilon = \bar{\varepsilon} \, (\eta, \, v), \quad \text{where} \quad v \equiv \frac{1}{\varrho} \, ,$$

$$\theta = \frac{\partial \bar{\varepsilon} \, (\eta, \, v)}{\partial \eta} \, ,$$

$$\boldsymbol{T} = - \frac{\partial \bar{\varepsilon} \, (\eta, \, v)}{\partial v} \, \boldsymbol{1} + \mathfrak{g}^{\mathrm{D}} (\eta, \, \mathrm{grad} \; \theta, \, v, \, \boldsymbol{D}) \, ,$$

$$\boldsymbol{h} = \bar{\boldsymbol{h}} \, (\eta, \, \mathrm{grad} \; \theta, \, v, \, \boldsymbol{D}) \, ,$$

where $\mathfrak{g}^{\mathrm{D}}$ and $\boldsymbol{h}$ are isotropic functions. Some distinction of effects appears, but still not the classical full separation.

COLEMAN & MIZEL then set up an expansion procedure about $\boldsymbol{D} = \boldsymbol{0}$ and $\mathrm{grad} \; \theta = \boldsymbol{0}$. They find that, with an error

$$O \left(\sqrt{\mathrm{tr} \, \boldsymbol{D}^2 + (\mathrm{grad} \; \theta)^2} \right) ,$$

the constitutive equations become

$$\boldsymbol{T}^{\mathrm{D}} = \lambda \, (\mathrm{tr} \, \boldsymbol{D}) \, \boldsymbol{1} + 2 \mu \, \boldsymbol{D} \, ,$$

$$\boldsymbol{h} = \varkappa \, \mathrm{grad} \; \theta \, .$$

These are the classical, fully separated equations for a fluid with linear viscosity and linear heat conduction, the expected result. What has been found is a *position* for the classical theory within more general concepts. I recall the assumptions that have been used:

1. The fundamental laws of mechanics and thermodynamics.
2. The principle of material frame-indifference.
3. The principle of equipresence.
4. Material symmetry (in this case, the material is a fluid).
5. Smoothness sufficient to permit an expansion.

In this case the classical separation of effects thus follows, in no disagreement with the principle of equipresence, for sufficiently symmetric materials sufficiently near to equilibrium. In particular, the separation is in part an expression of the smoothness of natural phenomena.

The principle of equipresence suggested itself in the course of my work on rarefied gases in 1949. In those days we did not have good mathematical apparatus for exploring continuum mechanics, and expansions were used in a routine way. Looking back at that work now, one may observe that the importance of isotropy and smoothness for getting the classical separation was seen there but was obscured by the attempt to treat a complicated new situation. The work of COLEMAN & MIZEL, attaching my earlier ideas firmly to modern general principles, fixes the classical theory within a generalization of modest scope and shows how to correct it.

Before going on to a more general concept of interaction between mechanical and thermal phenomena, I remark upon extension of the same ideas to problems of diffusion. In 1957 I set up general equations expressing the interchange of mass, momentum, and energy among the constituents of an interdiffusing, chemically reacting mixture, but my treatment of entropy, suffering (though under protest) from the influence of the organized linear literature on "irreversible processes", employed equations of state from the outset. In recent weeks Mr. BOWEN, of this university, has remedied this defect by establishing a corresponding framework for the entropy of a mixture, in the same generality as the previously known equations of change for mass, linear momentum, and internal energy. On the basis of this framework he has set up for study a theory of mixtures of visco-elastic substances of the type previously studied by COLEMAN & MIZEL for a single medium. Adopting the principle of equipresence, he allows the field variables for all constituents to appear as arguments in the constitutive equations for any one. He shows that consistency with the principle of growth of entropy restricts the way in which the equilibrium part of the stress for one constituent may depend on the quantities associated with the other constituents. Earlier GREEN & ADKINS in an attempt to formulate a rational general theory of diffusion had objected to the principle of equipresence because it seemed to render the constituents indistinguishable from one another. Mr. BOWEN'S work shows that their objection is not just, since in fact the relation giving the equilibrium portion of the stress in terms of the deformation for each constituent must remain in a mixture exactly the same as it is in a pure specimen.

The papers of COLEMAN & NOLL and of COLEMAN & MIZEL, as well as showing that the position based on equipresence and the Clausius-Duhem inequality is tenable, scouted the scene for the general attack on the thermodynamics of simple materials. A comprehensive theory has recently been achieved by Mr. COLEMAN. He supposes that the four main fields of mechanics and thermodynamics are determined by the histories of deformation and temperature, the constitutive functionals being allowed to depend on grad θ as a parameter:

$$T = \mathop{\mathfrak{T}}_{s=0}^{\infty} \left(F(t-s),\ \theta(t-s)\,;\ \mathrm{grad}\ \theta \right),$$

$$h = \mathop{\mathfrak{H}}_{s=0}^{\infty} \left(F(t-s),\ \theta(t-s)\,;\ \mathrm{grad}\ \theta \right),$$

$$\varepsilon = \mathop{e}_{s=0}^{\infty} \left(F(t-s),\ \theta(t-s)\,;\ \mathrm{grad}\ \theta \right),$$

$$\eta = \mathop{\mathfrak{h}}_{s=0}^{\infty} \left(F(t-s),\ \theta(t-s)\,;\ \mathrm{grad}\ \theta \right).$$

These equations reduce to those of classical thermostatics when the functionals reduce to functions obeying certain relations. In general, they replace both thermostatics and the visco-elastic thermodynamics of COLEMAN & MIZEL by a true theory of thermoenergetic processes changing in time, with all corresponding effects of memory. More mathematics is needed in the working, but to develop this broader theory the basic line of thought is much the same. It is proved that grad θ must drop out of the equations for ε and η, and that $\mathfrak{h}$ and $\mathfrak{T}$ are determined from e by functional differentiation, generalizing the classical forms of the stress-strain and temperature-entropy relations. The expansion procedure then introduced rests on the principle of fading memory. COLEMAN finds that the classical thermostatic theory results as a first approximation either for a very slow or for a very fast deformation. At this level of generality, "thermostatics" really means the classical theory of finite elastic strain. This theory, recently the subject of a great revival, thus appears at four different positions in modern continuum mechanics:

1. As the general theory of simple materials in equilibrium.

2. As a theory of special materials in all deformation processes.

3. As the common approximation for all simple materials with fading memory if deformed very slowly.

4. As the common approximation for all simple materials with fading memory if deformed very fast.

The first two cases are obvious and the third was generally believed. Some years ago a number of persons began to conjecture the fourth case, and it was encouraging, after my first lecture at Johns Hopkins in 1961, to learn that some of the formulae for purely elastic waves I had presented in it Mr. BELL found immediately consistent with his long-standing measurements on high velocity impact of bars. It is obvious that pure elasticity (at least as that theory is usually interpreted) does not describe the entire deformation in his experiment, which results in large permanent set; it is not obvious, but it may be true, that the onset of the deformation, when the waves begin to travel and the measurements are made, may be purely elastic in nature. COLEMAN'S theorem makes this agreement, which puzzled me in 1961, now seem natural.

In fact, COLEMAN & GURTIN in their great analysis of wave motions, now just going to press, have proved that in any simple material with fading memory, the laws of propagation of acceleration waves, at a given place and time, are the same as those in an appropriately defined elastic material. The nature of that elastic material varies according to the deformation history, but any property independent of the particular form of the stress-strain relation in elasticity will carry over unchanged. Thus the consistency of Mr. BELL'S results on waves with universal relations from elasticity, despite the large permanent set resulting, can be understood easily. (Of course, I am *not* saying that the materials on which Mr. BELL experiments are described by this theory or by any theory. Statements of this sort must be left to experimenters. All I am saying is that the agreement found by Mr. BELL need no longer puzzle the theorist, since it is *of the type* now included by at least one class of simple theories.)

All of the classic assertions about the directions of special processes find their places as clearly stated and mathematically proved theorems in Mr. COLEMAN'S system. Writing $\theta\sigma$ for the first Fréchet derivative of the free energy with respect to the deformation-temperature history $\big(F(t-s),\ \theta(t-s)\big)$, evaluated at the present deformation-temperature $\big(F(t),\ \theta(t)\big)$, COLEMAN calls σ the *internal dissipation* and proves, as a part of his fundamental theorem, that

$$\sigma \geqq 0.$$

Moreover,

$$\boldsymbol{h} \cdot \operatorname{grad} \theta \geqq -\varrho\, \theta^2\, \sigma.$$

These are the fundamental inequalities of the thermodynamics of deforming simple materials. As a corollary, COLEMAN has proved a definitive theorem on thermodynamic extremes: Of all processes ending with given values of the deformation

$$F \text{ and the } \begin{Bmatrix} \text{temperature} \\ \text{internal energy} \\ \text{entropy} \end{Bmatrix},$$

that corresponding to constant values of

$$F \text{ and } \begin{Bmatrix} \theta \\ \varepsilon \\ \eta \end{Bmatrix} \text{ for all times has } \begin{Bmatrix} \text{the least free energy.} \\ \text{the greatest entropy.} \\ \text{the least internal energy.} \end{Bmatrix}$$

He has proved that in every isothermal, iso-energetic, or isentropic process starting from a state of equilibrium, the total stress work done in traversing a closed path is non-negative. He has derived also inequalities for the rates of change of the thermodynamic potentials:

$$\dot{\psi} \leqq \frac{1}{\varrho} \operatorname{tr}(\boldsymbol{T}\boldsymbol{D}) - \eta \dot{\theta},$$

$$\theta \dot{\eta} \geqq \dot{\varepsilon} - \frac{1}{\varrho} \operatorname{tr}(\boldsymbol{T}\boldsymbol{D}),$$

$$\theta \dot{\eta} \geqq q + \frac{1}{\varrho} \operatorname{div} \boldsymbol{h}.$$

Of course, results of similar form were familiar in the standard books, but they referred to extremal definitions of equilibrium, to materials obeying a caloric equation of state, or to "quasi-static processes", in various combinations. Here, in infinitely greater generality, they refer to simple materials with long-range memory and to arbitrary changes of deformation and temperature. More important, they are *theorems*, *proved* to follow from the principle of fading memory and the Clausius-Duhem inequality.

A century of talk about irreversible processes has been replaced during the past year by a definite theory of temperature and entropy changes in deforming materials. As is usual in sound new work, the classical formulae turn out to be neither wrong nor sufficient, merely special and approximate.

Often I am asked what relation the work of COLEMAN and his collaborators bears to the official "thermodynamics of irreversible processes", conveniently denoted by the term "Onsagerist[1]". I am glad to say, none whatever! It is high time to prick the bubble of Onsagerism. In the halls of the Johns Hopkins University, where first in the Western Hemisphere theoretical physics was given a place at the table, I hope to be allowed a few plain words.

The cultivators of the linear "thermodynamics of irreversible processes" call upon a small perturbation of thermostatics. Striving to shore up the collapsing illusion that energy is everything, they have stretched interpretation of results ground in the old way out of equations of state to cover a few new corners of science by smothering the subject in a blanket of linear and

---

[1] The term "Onsagerist" is intended to divorce my statements here from any personal reference to L. ONSAGER, who, so far as I know, after his interesting statistical essay of 1931 has stood aloof from the colossal discharge of literature on "thermodynamics of irreversible processes".

symmetric mud. Refusing to recognize that motions are produced by forces, they have treated the laws of mechanics like family skeletons; refusing to recognize that material properties are expressed by constitutive equations subject to principles of invariance, they have excused their blank assumptions by appeal to a non-existent theorem of algebra and to approximations alleged to follow from the kinetic theory of gases and statistical mechanics; refusing to recognize the role of mathematics in physical science, they have written whole books in which not a single non-trivial problem is stated or solved in mathematical terms. Their science begins and ends[1] with the equation

$$x = A X, \quad \text{where } x \text{ and } X \text{ are any quantities such that} \quad \dot{\eta} \propto x \cdot X,$$

and where what is variously called ONSAGER'S "theorem", "principle", "law", "postulate", or "axiom", asserts that the tensor A, which converts "forces" (or "fluxes") X into "fluxes" (or "forces") x, is always symmetric in purely mechanical problems, while its skew part is subject to a rule of disparity when electromagnetic phenomena are taken into account. Whenever in mechanics a tensor occurs that is symmetric, someone rises to claim it as obedient to "ONSAGER'S reciprocal relations".

I have not time today to chart the rotten foundation on which the Onsagerist school has built. Even were the basis sound, any reader of an Onsagerist book can see three characterizing flaws:

1. Forces and fluxes are not defined as physical quantities. Rather, they are universals, which the author selects afresh in each illustrative case. The only selection rules the Onsagerists

---

[1] The summary must be qualified in the case of systems of a *finite* number of degrees of freedom. There the "forces" and "fluxes", according to CASIMIR'S rule, are to be taken as conjugate variables, and the resulting theory is self-consistent. When fields are considered, no counterpart of this rule is known. As Mr. COLEMAN and I have shown, under free choice of "forces" and "fluxes" it is always, and trivially, possible to select them so as to make A symmetric or unsymmetric, as we please, so that "ONSAGER's theorem" loses all phenomenological content.

Of course, some matrices of physical data *are* symmetric. Unless the Onsagerists can show some rule for selecting forces and fluxes (as they do for finite systems but not for continuous ones), such facts of experiment indicate no more than a happy chance in the choice of variables.

give us involve such concepts as "reversal of the velocities of the molecules", while the experimenter measuring the work done or the deformation effected or the mass diffused in a specimen of a material is scarcely in a position to consult the molecules as to what effect they could produce by going backward. Even were Onsagerist thermodynamics self-consistent, it is not effective. There is no physics in the structure, which is merely a trivial part of linear algebra with new names.

2. Since the general principles of the rest of physics are not brought to bear, no flows or deformations of substances can be calculated, except by slipping in some special and often unstated assumption, which in many cases is enough to solve the special problem without appeal to "Onsagerist" principles in the first place. The structure is not raised to the height of mathematics.

3. On questions or large deformation the Onsagerists are silent, for to speak of a non-linear constitutive equation seems to be forbidden by their very vocabulary. (The one exceptional Onsagerist paper proves the rule, for in it the whole structure has to be recast, only to get as end result no more than a drastic specialization within an old non-linear theory of fluids which experiment and rational theory alike had shown long before to be too special as it was.)

What I have just said is not new. Rumblings of this kind have been heard ever since the appearance of the first Onsagerist monograph in 1947, but up to now, the critics have been silenced by the questions, "Can you do any better with these phenomena?" Because of COLEMAN's work, we can now answer, Yes! The variables used by COLEMAN are not universals but rather the quantities already familiar in mechanics: motion, stress, heat flux, energy, temperature, entropy. He proposes no new physical laws but rather, after the pattern of the great field theories of the past, finds unequivocal, general statements of the old ones and shows how to interpret and apply them, and them only, to materials of various kinds. Above all the contrast with the Onsagerists comes in use of *mathematics* in its traditional power, to extract logically from a few simple and plausible assumptions a wealth of consequences which before were unsuspected or at best imperfectly guessed. The basic principles apply equally to deformations of any magnitude and to materials of arbitrary symmetry. More special results follow, as they should, when particular material symmetries such as isotropy are laid down, or when we consider particular classes of deformations such as viscometric flows or retarded flows or impulses.

COLEMAN'S theory does not pretend to cover the range of physical phenomena claimed by the Onsagerists. After all, the fields basic in continuum mechanics are finite in number: motion, stress, energy, heat flux, temperature, entropy, charge-current, magnetic flux, polarization, magnetization — perhaps a few more. The generality of "forces" and "fluxes" is both unnecessary for this range and physically empty if it extends further. Continuum physics seeks a correct theory of *just these fields*, not x and X. COLEMAN'S theory covers the first six, with no attempt to include electromagnetic effects. BOWEN'S theory makes a start on mixtures. Different ideas will be needed for the thermodynamics of electromagnetism, but they will be thermo-electromagnetic ideas, not x and X or names for the entries in a matrix.

Like any other theory, COLEMAN'S will be found to have limitations in practice, to apply to some ranges of phenomena but fail to apply to others. Such a sharing between success and failure in application is common to all true scientific theories. A theory that covers everything, that cannot be contradicted by experiment, that requires only proper interpretation to be universal, belongs not to science but to religion.

IV. Electrified materials

In a Euclidean frame which is also a Lorentz frame, MAXWELL'S field equations are

$$\text{div}\, \boldsymbol{J} + \frac{\partial Q}{\partial t} = 0 \quad \text{(conservation of charge-current)},$$

$$\text{curl}\, \boldsymbol{E} + \frac{\partial \boldsymbol{B}}{\partial t} = \boldsymbol{0} \quad \text{(FARADAY's law of induction)},$$

$$\text{div}\, \boldsymbol{B} = 0 \quad \text{(conservation of magnetic flux)},$$

and two of MAXWELL's potential relations are

$$Q = \text{div}\, \boldsymbol{D},$$

$$\boldsymbol{J} = \text{curl}\, \boldsymbol{H} - \frac{\partial \boldsymbol{D}}{\partial t}.$$

Here $\boldsymbol{J}$ is the current density, Q is the charge density, $\boldsymbol{E}$ is the electric field, $\boldsymbol{B}$ is the magnetic flux density, $\boldsymbol{D}$ is the charge potential, and $\boldsymbol{H}$ is the current potential. The Maxwell-Lorentz aether relations assert that we may choose the potentials $\boldsymbol{D}$ and $\boldsymbol{H}$ in such a way that

$$\boldsymbol{D} = \varepsilon_0 \boldsymbol{E}, \qquad \boldsymbol{H} = \frac{1}{\mu_0} \boldsymbol{B},$$

where ε_0 and μ_0 are dimensional constants. These equations hold everywhere, within matter and without it. The action of an electromagnetic field upon matter gives rise to special kinds of charge and current, which are conveniently decsribed in terms of the magnetization $\boldsymbol{M}$ and the polarization $\boldsymbol{P}$. In a stationary body devoid of free charge and free current,

$$Q = -\,\text{div}\, \boldsymbol{P}, \qquad \boldsymbol{J} = \frac{\partial \boldsymbol{P}}{\partial t} + \text{curl}\, \boldsymbol{M}.$$

Accordingly, if we write $\mathfrak{D}$ and $\mathfrak{H}$ for the partial potentials, so that

$$\mathfrak{D} \equiv \boldsymbol{D} + \boldsymbol{P}, \qquad \mathfrak{H} \equiv \boldsymbol{H} - \boldsymbol{M},$$

$\mathfrak{D}$ and $\mathfrak{H}$ being called the potentials of "free charge" and "free current", then the field equations and potential relations become

$$\frac{\partial \boldsymbol{B}}{\partial t} + \text{curl } \boldsymbol{E} = \boldsymbol{0},$$

$$\text{div } \boldsymbol{B} = 0,$$

$$\text{curl } \mathfrak{H} - \frac{\partial \mathfrak{D}}{\partial t} = \boldsymbol{0},$$

$$\text{div } \mathfrak{D} = 0,$$

and these partial field equations we take as our starting point.

An electromagnetic wave (of second order) is a surface across which the fields are continuous, but their gradients and time rates may suffer smooth jump discontinuities: $[\text{grad } \boldsymbol{E}]$, $[\partial \boldsymbol{E}/\partial t]$, *etc.* According to the compatibility conditions of MAXWELL, WEINGARTEN, and HADAMARD, for such a wave to exist and persist it is necessary that there exist a vector $\boldsymbol{e}$ and a scalar u such that

$$[\text{div } \boldsymbol{E}] = \boldsymbol{n} \cdot \boldsymbol{e}, \quad [\text{curl } \boldsymbol{E}] = \boldsymbol{n} \times \boldsymbol{e}, \quad \left[\frac{\partial \boldsymbol{E}}{\partial t}\right] = -u\,\boldsymbol{e},$$

where $\boldsymbol{n}$ is the unit normal to the wave front. The vector $\boldsymbol{e}$ is the electric amplitude of the wave, and the scalar u is its normal speed of advance in space. The jumps in the derivatives of the three remaining fields satisfy similar conditions of compatibility, *e.g.*,

$$\left[\frac{\partial \boldsymbol{B}}{\partial t}\right] = -u\,\boldsymbol{b}, \quad [\text{div } \boldsymbol{B}] = \boldsymbol{n} \cdot \boldsymbol{b},$$

$$\left[\frac{\partial \mathfrak{D}}{\partial t}\right] = -u\,\boldsymbol{d}, \quad [\text{div } \mathfrak{D}] = \boldsymbol{n} \cdot \boldsymbol{d},$$

$$[\text{curl } \mathfrak{H}] = \boldsymbol{n} \times \boldsymbol{h}.$$

The vectors $\boldsymbol{b}$, $\boldsymbol{d}$, and $\boldsymbol{h}$ are the amplitudes of the wave in magnetic flux, potential of free charge, and potential of free current.

Calculating the jumps of the left-hand sides of the field equations across the wave, by use of the conditions of compatibility we obtain

$$-u\,\boldsymbol{b} + \boldsymbol{n} \times \boldsymbol{e} = \boldsymbol{0},$$

$$\boldsymbol{n} \cdot \boldsymbol{b} = 0,$$

$$\boldsymbol{n} \times \boldsymbol{h} + u\,\boldsymbol{d} = \boldsymbol{0},$$

$$\boldsymbol{n} \cdot \boldsymbol{d} = 0.$$

The second and fourth of these equations, asserting that the wave is transverse with respect to magnetic flux and potential of free charge, we may discard because they follow from the first and third equations, respectively, provided only that $u \neq 0$: The wave is actually moving, as we shall assume henceforth. Thus the definitive equations for propagation of second-order waves are

$$-u\,\boldsymbol{b}+\boldsymbol{n}\times\boldsymbol{e}=0\,,$$

$$\boldsymbol{n}\times\boldsymbol{h}+u\,\boldsymbol{d}=0\,.$$

These equations, like our starting equations themselves, are valid in any body which is at rest in a Euclidean-Lorentz frame and is devoid of free charge and free current.

In the luminiferous aether, $\boldsymbol{M}=0$ and $\boldsymbol{P}=0$. Thus

$$\mathfrak{H}=\frac{1}{\mu_0}\,\boldsymbol{B}\,,\qquad \mathfrak{D}=\varepsilon_0\,\boldsymbol{E}\,.$$

It is not hard to prove that as a consequence, the amplitudes of the waves satisfy the same relations:

$$\boldsymbol{b}=\mu_0\,\boldsymbol{h}\,,\qquad \boldsymbol{d}=\varepsilon_0\,\boldsymbol{e}\,,$$

showing that the wave is entirely transverse. Substitution into the conditions of compatibility yields

$$-u\,\mu_0\,\boldsymbol{h}+\boldsymbol{n}\times\boldsymbol{e}=0\,,$$

$$\boldsymbol{n}\times\boldsymbol{h}+u\,\varepsilon_0\,\boldsymbol{e}=0\,.$$

Substituting the second equation into the first and using the fact that $\boldsymbol{n}\times(\boldsymbol{n}\times\boldsymbol{h})=-\boldsymbol{h}$ gives the propagation condition:

$$\left(-\frac{\mu_0\,\varepsilon_0}{s^2}+1\right)\boldsymbol{e}=0\,,$$

where

$$s=1/u\,,\quad \text{the } \textit{slowness} \text{ of the wave.}$$

We have shown, then, that waves in the aether are entirely transverse, and, provided they travel with slowness $s=\sqrt{\mu_0\,\varepsilon_0}$, they may have any transverse amplitude whatever.

While it seems to be right for the aether, this type of restriction upon waves is extraordinary for other materials. In elastic solids, for example, kinematic waves of at least two kinds are possible, and they travel at different speeds, determined by FRESNEL'S ellipsoid, which was introduced to describe refraction in crystals. Various special theories have been advanced for other particular optical effects observed in matter. According to the Faraday effect, right and left circularly polarized waves travelling down the lines of induction in an isotropic material travel with different speeds, and the plane of polarization of a linearly polarized wave rotates at a rate proportional to the distance travelled. According to the Voigt effect, light propagating normally to a strong magnetic field suffers double refraction. According to the Kerr effect, double refraction occurs also when light propagates down the lines of force in a strong electric field. These effects are not included in the classical theory of light in the aether, which is fully linear. While special theories, based on minor perturbations of the aether theory, have been advanced by physicists to account for these particular effects, TOUPIN & RIVLIN in 1961 were the first to apply the methods of modern non-linear continuum mechanics to the electromagnetic field in transparent materials. Their results can be simplified, and it is with a view to this end that I have presented the classical theory of waves in the aether in a form fit for immediate generalization. In the aether, waves carrying jumps in the fields themselves, not merely their gradients, obey exactly the same laws. Such waves are the analogues of shock waves in mechanics, while waves of second or higher order are the analogues of sound waves. Since the aether is a fully linear medium, all kinds of waves have common laws of propagation. In the stationary Maxwellian dielectric, defined by the constitutive equations

$$P = \varepsilon_0 \chi E, \qquad M = 0,$$

or equivalently,

$$\mathfrak{D} = \varepsilon E, \qquad \varepsilon = \varepsilon_0 (1 + \chi), \qquad \mathfrak{H} = H = \frac{1}{\mu_0} B,$$

the same result holds. However, in a non-linear medium such as an ordinary gas or finitely elastic solid, shock waves are more elaborate and various than sound waves, and in the same way, if we seek to explain wave propagation in non-linear electromagnetism, it is natural to begin with waves of second order rather than first.

The aether pervades all matter, and its properties are unaffected by its circumstances. Matter itself, however, reacts to the application of a non-vanishing electromagnetic field. For example, light propagates differently when passing through a material. As the simplest case to consider, let the magnetization and polarization be determined, uniquely and instantaneously, by the applied fields:

$$P = P(E, B),$$

$$M = M(E, B).$$

More general response, in which the material accumulates the effects of fields applied in the past, is easy to conjecture, but we first propose for study this simplest possibility of constitutive equations for *a stationary, rigid, transparent material*, devoid of free charge and free current. Equivalently,

$$\mathfrak{D} = \mathfrak{D}(E, B),$$

$$\mathfrak{H} = \mathfrak{H}(E, B).$$

Constitutive equations of this kind where $\mathfrak{D}$ and $\mathfrak{H}$ are linear functions were proposed by B. D. H. TELLEGEN in 1948 and studied for isotropic materials by LL. G. CHAMBERS in 1955. They found some but not all of the effects we shall now derive. TOUPIN & RIVLIN proposed a more general theory in which $\mathfrak{D}$ and $\mathfrak{H}$ may depend upon the entire history of the fields E and B, but these latter are supposed to remain nearly constant in time.

The constitutive equations imply linear relations among the wave amplitudes. *E.g.*, since

$$\frac{\partial \mathfrak{D}}{\partial t} = \frac{\partial \mathfrak{D}}{\partial E} \frac{\partial E}{\partial t} + \frac{\partial \mathfrak{D}}{\partial B} \frac{\partial B}{dt},$$

it follows that

$$d = \frac{\partial \mathfrak{D}}{\partial E} e + \frac{\partial \mathfrak{D}}{\partial B} b;$$

likewise

$$h = \frac{\partial \mathfrak{H}}{\partial E} e + \frac{\partial \mathfrak{H}}{\partial B} b.$$

In the more general theory of TOUPIN & RIVLIN,

$$d = \Phi e + \Psi b,$$

$$h = \Omega e + \Lambda b,$$

where the four tensors $\boldsymbol{\Phi}$, $\boldsymbol{\Psi}$, $\boldsymbol{\Omega}$, $\boldsymbol{\Lambda}$ are not necessarily derivable from vector functions as they are here:

$$\boldsymbol{\Phi}=\frac{\partial\mathfrak{D}}{\partial\boldsymbol{E}}, \quad \boldsymbol{\Psi}=\frac{\partial\mathfrak{D}}{\partial\boldsymbol{B}}, \quad \boldsymbol{\Omega}=\frac{\partial\mathfrak{H}}{\partial\boldsymbol{E}}, \quad \boldsymbol{\Lambda}=\frac{\partial\mathfrak{H}}{\partial\boldsymbol{B}}.$$

Substitution in the compatibility conditions now yields

$$-u\,\boldsymbol{b}+\boldsymbol{n}\times\boldsymbol{e}=0,$$

$$\boldsymbol{n}\times(\boldsymbol{\Omega}\,\boldsymbol{e}+\boldsymbol{\Lambda}\,\boldsymbol{b})+u\,(\boldsymbol{\Phi}\,\boldsymbol{e}+\boldsymbol{\Psi}\,\boldsymbol{b})=0.$$

We can use the first of these equations to eliminate $\boldsymbol{b}$ altogether, since we have assumed that $u\neq0$, so the singular surface is not stationary:

$$\boldsymbol{n}\times(\boldsymbol{\Omega}\,\boldsymbol{e})+\frac{1}{u}\,\boldsymbol{n}\times\boldsymbol{\Lambda}\,(\boldsymbol{n}\times\boldsymbol{e})+u\,\boldsymbol{\Phi}\,\boldsymbol{e}+\boldsymbol{\Psi}\,(\boldsymbol{n}\times\boldsymbol{e})=0.$$

This is the *amplitude* condition, expressed entirely in terms of the electric amplitude $\boldsymbol{e}$. We may write it in the form

$$\boldsymbol{\chi}(\boldsymbol{n})\,\boldsymbol{e}=0,$$

where $\boldsymbol{\chi}(\boldsymbol{n})$, the *optical tensor*, has the following Cartesian components:

$$\chi_{ik}=\Phi_{ik}+s\,(e_{ijr}\,n_j\,\Omega_{rk}+\Psi_{il}\,e_{ljk}\,n_j)+s^2 e_{ijs}\,n_j\,\Lambda_{sq}\,e_{qrk}\,n_r.$$

This is Toupin & Rivlin's main result.

A non-zero electric amplitude $\boldsymbol{e}$ such as to satisfy the amplitude condition exists if and only if

$$\det\boldsymbol{\chi}(\boldsymbol{n})=0.$$

This is the *propagation condition*, which determines the slowness s at which a wave normal to $\boldsymbol{n}$ may propagate in a stationary transparent medium subjected to a given electromagnetic field $\boldsymbol{E}$ and magnetic flux density $\boldsymbol{B}$. If s satisfies the propagation condition, the direction of the corresponding amplitude $\boldsymbol{e}$ is determined from the equation $\boldsymbol{\chi}(\boldsymbol{n})\,\boldsymbol{e}=0$. Of course the magnitude of the amplitude is arbitrary. While the propagation condition may seem to be an equation of degree six, it is in fact of degree four. To see this, note that

$$\boldsymbol{\chi}(\boldsymbol{n})\,\boldsymbol{n}=\boldsymbol{\Phi}\boldsymbol{n}+s\,\boldsymbol{n}\times\boldsymbol{\Omega}\,\boldsymbol{n},$$

$$\boldsymbol{n}\cdot\boldsymbol{\chi}(\boldsymbol{n})\,\boldsymbol{n}=\boldsymbol{n}\cdot\boldsymbol{\Phi}\boldsymbol{n}.$$

Since this vector and scalar are of degrees 1 and 0 in s, inspection of det χ written in terms of an orthogonal basis, one vector of which is n, shows that det χ is of degree at most four in s. Since the equation det $\chi = 0$ is a quartic with real coefficients, its roots occur in complex-conjugate pairs. Therefore, in general, *in a stationary transparent medium there are two kinds of electromagnetic waves.* The phenomenon of double refraction is thus seen to be an ordinary effect of transparency. Of course, it may fail to manifest itself in special cases, when the two kinds of waves coalesce to one. For example, the optical tensor of the aether is

$$\chi(n) = \left(-\frac{s^2}{\mu_0} + \varepsilon_0 \right) 1 + \frac{s^2}{\mu_0} \, n \otimes n \,,$$

so that there are in any direction n two identical pairs of waves, with slowness $\pm\sqrt{\varepsilon_0\mu_0}$, independent of n. In general, however, preferred directions are determined by the existing fields E and B, and the waves do not propagate at the same speeds even in opposite senses of the same direction.

That the vibrations of the aether are of a kind different from those of an unstressed isotropic linearly elastic material is well known. While in the elastic material there are two kinds of waves, transverse and longitudinal, the aether allows only transverse waves. We are now able to make a similar comparison between anisotropic materials, transparent on the one hand or elastic on the other. In a pre-stressed elastic material there are in general three kinds of waves, with mutually orthogonal amplitudes, and both amplitudes and speeds are, subject to certain assumptions, real. In a transparent medium affected by an electromagnetic field, there are only two kinds of waves, and both amplitudes and speeds, are, in general, complex.

Consider now an isotropic material, defined as one such that the isotropy groups of $\mathfrak{D}$ and $\mathfrak{H}$ are the full orthogonal group. That is, if $\sigma = \det Q$, then for all orthogonal Q

$$\mathfrak{D}(Q E, \sigma Q B) = Q \mathfrak{D}(E, B) \,,$$

$$\mathfrak{H}(Q E, \sigma Q B) = \sigma Q \mathfrak{H}(E, B) \,,$$

where we are taking account of the fact that B and $\mathfrak{H}$ are axial vectors. The general solutions of these functional equations can be found. It is easier to begin with the more general problem of finding a representation for a hemitropic vector-valued function $\boldsymbol{f}$ of two vectors $\boldsymbol{K}$ and $\boldsymbol{L}$:

$$\boldsymbol{f}(\boldsymbol{QK},\ \boldsymbol{QL}) = \boldsymbol{Qf}(\boldsymbol{K},\ \boldsymbol{L})$$

for all proper orthogonal $\boldsymbol{Q}$. For such a function,

$$\boldsymbol{f}(\boldsymbol{K},\ \boldsymbol{L}) \cdot \boldsymbol{M}$$

is a hemitropic scalar function of the three vectors $\boldsymbol{K}$, $\boldsymbol{L}$, $\boldsymbol{M}$, for an arbitrary vector $\boldsymbol{M}$. By a theorem of CAUCHY, therefore,

$$\boldsymbol{f}(\boldsymbol{K},\ \boldsymbol{L}) \cdot \boldsymbol{M} = g\,(K^2,\ L^2,\ M^2,\ \boldsymbol{K} \cdot \boldsymbol{L},\ \boldsymbol{K} \cdot \boldsymbol{M},\ \boldsymbol{L} \cdot \boldsymbol{M},\ \boldsymbol{K} \times \boldsymbol{L} \cdot \boldsymbol{M}).$$

But $\boldsymbol{f}(\boldsymbol{K},\ \boldsymbol{L}) \cdot \boldsymbol{M}$ is linear and homogeneous in $\boldsymbol{M}$:

$$\boldsymbol{f}(\boldsymbol{K},\ \boldsymbol{L}) \cdot \boldsymbol{M} = [f_1 \boldsymbol{K} + f_2 \boldsymbol{L} + f_3 (\boldsymbol{K} \times \boldsymbol{L})] \cdot \boldsymbol{M},$$

whence it follows that

$$\boldsymbol{f}(\boldsymbol{K},\ \boldsymbol{L}) = f_1 \boldsymbol{K} + f_2 \boldsymbol{L} + f_3 (\boldsymbol{K} \times \boldsymbol{L}),$$

where the scalars f_i are functions of K^2, L^2, and $\boldsymbol{K} \cdot \boldsymbol{L}$.

Accordingly, in a hemitropic material both $\mathfrak{D}$ and $\mathfrak{H}$ have representations of the form just derived. In an isotropic material, somewhat more explicit representations are valid. Recall that $\mathfrak{D}$ and $\boldsymbol{E}$ are polar vectors, while $\boldsymbol{B}$ and $\mathfrak{H}$ are axial vectors. That is, under a central inversion $\mathfrak{D}$ and $\boldsymbol{E}$ change into $-\mathfrak{D}$ and $-\boldsymbol{E}$, while $\boldsymbol{B}$ and $\mathfrak{H}$ remain invariant. Both $\boldsymbol{B} \times \boldsymbol{E}$ and $\boldsymbol{B} \cdot \boldsymbol{E}$ are inverted. The requirement that the representations for $\mathfrak{D}$ and $\mathfrak{H}$ be invariant under the central inversion thus leads to the following reduced forms:

$$\mathfrak{D} = \delta_1 \boldsymbol{E} + \delta_2 (\boldsymbol{B} \cdot \boldsymbol{E})\,\boldsymbol{B} + \delta_3 \boldsymbol{B} \times \boldsymbol{E},$$

and

$$\mathfrak{H} = \eta_1 (\boldsymbol{B} \cdot \boldsymbol{E})\,\boldsymbol{E} + \eta_2 \boldsymbol{B} + \eta_3 \varepsilon\,(\boldsymbol{B} \cdot \boldsymbol{E})\,(\boldsymbol{B} \times \boldsymbol{E}),$$

where the coefficients δ_i and η_i are functions of the three scalars

$$E^2,\ B^2,\ (\boldsymbol{B} \cdot \boldsymbol{E})^2,$$

or, if we prefer, are *even* scalar functions of $\boldsymbol{B}$ and $\boldsymbol{E}$. Substitution of these formulae in the general expression for the optical tensor yields a rather complicated expression in terms of the coefficient-functions δ_i and η_i and their partial derivatives, three apiece, making 24 optical coefficients of an isotropic transparent medium, all of which are functions of the applied fields B and E. In the more general theory of TOUPIN & RIVLIN, the isotropic material may have as many as 36 independent optical coefficients. To obtain their result, we need only assume that the tensors Φ, Ψ, Λ, Ω are isotropic functions of B and E.

TOUPIN & RIVLIN worked with the theory of infinitesimal oscillations rather than second-order singular surfaces. The results are of the same form, but complex amplitudes and slownesses have immediate application. When $E = 0$ and $n \parallel B$, the Faraday effect is described. When $E = 0$ and $n \perp B$, the Voigt effect occurs. Other simple cases are obtained for other special types of fields and waves. All the previously known effects in transparent media are found as special cases within an embracing theory, and various new effects are predicted. While the individual results, new and old, are interesting in themselves, the connecting of the various special theories, each at its proper place within a unified theory of optical phenomena in stationary media, is more important for understanding electromagnetism as a branch of natural philosophy. Perhaps the difference between the old and new ways of looking at these phenomena can be explained best by a simple summary. In the older electromagnetic theory, these effects could not occur according to the general equations, and to account for each in turn some special alteration had to be made at some point or other. In the present view, all these effects and many more arise automatically, and what seems surprizing is that any theory of electromagnetism in matter can ever have overlooked them. Their presence does not support any particular theory of electrified materials; rather, it is typical of all of them, the differences being of degree rather than kind. In the older view, double refraction was an anomaly which cried for explanation; today, MACCULLAGH and MAXWELL seem ingenious in having been able to construct aether theories that avoid it.

I have selected TOUPIN & RIVLIN's theory of electro-magneto-optical effects as the first example because each step is so easy and natural for anyone equipped with the mathematical tools and the approach learned from modern researches in rational mechanics. Problems of electromechanical interaction are more difficult, since the concepts of electromagnetism and of mechanics do not fit together perfectly. A particular theory, important as being the first to consider finite deformations correctly, was proposed by Mr. TOUPIN in 1956 and generalized in 1963. This is the theory of the elastic dielectric.

The dielectric is supposed to be at rest, or nearly so, with respect to the aether. TOUPIN employs following concepts from mechanics:

ϱ	mass density
$\dot{\boldsymbol{x}}, \ddot{\boldsymbol{x}}$	velocity, acceleration
$\boldsymbol{F}$	deformation gradient
$\boldsymbol{T}$	stress tensor
ε	specific internal energy

and the following from electromagnetism beyond those used before:

$$\mathfrak{E} = \boldsymbol{E} + \dot{\boldsymbol{x}} \times \boldsymbol{B} \quad \text{electromotive intensity}.$$

Also, to the extent that motion is considered, the relation between $\mathfrak{H}$ and $\boldsymbol{B}$ must be generalized:

$$\mathfrak{H} = \frac{1}{\mu_0} \boldsymbol{B} + \dot{\boldsymbol{x}} \times \boldsymbol{P}.$$

The basic assumption of the theory is that deformation and polarization determine the energy, the stress, and the electromotive intensity:

$$\varepsilon = \hat{\varepsilon}(\boldsymbol{F}, \boldsymbol{P}),$$

$$\boldsymbol{T} = \mathfrak{g}(\boldsymbol{F}, \boldsymbol{P}) = \boldsymbol{T}^T,$$

$$\mathfrak{E} = \hat{\mathfrak{E}}(\boldsymbol{F}, \boldsymbol{P}).$$

The principle of equipresence is satisfied. In accord with the views of LORENTZ, $-$ div $\boldsymbol{P}$ is a charge density, and

$$\frac{\partial \boldsymbol{P}}{\partial t} - \operatorname{curl}(\boldsymbol{P} \times \dot{\boldsymbol{x}})$$

is the corresponding current density. Thus if $\boldsymbol{P}^*$ is written for the convected time flux of $\boldsymbol{P}$:

$$\boldsymbol{P}^* \equiv \frac{\partial \boldsymbol{P}}{\partial t} + \dot{\boldsymbol{x}} \operatorname{div} \boldsymbol{P} + \operatorname{curl}(\boldsymbol{P} \times \dot{\boldsymbol{x}}),$$

there is an electromagnetic body force or "Lorentz force"

$$\varrho \boldsymbol{b} = - \mathfrak{E} \operatorname{div} \boldsymbol{P} + \boldsymbol{P}^* \times \boldsymbol{B}$$

and a corresponding heat absorption

$$\varrho q = \boldsymbol{P}^* \cdot \dot{\boldsymbol{\mathfrak{E}}}.$$

Accordingly, the equations of linear momentum and energy are

$$\text{div } \boldsymbol{T} - \boldsymbol{\mathfrak{E}} \text{ div } \boldsymbol{P} + \boldsymbol{P}^* \times \boldsymbol{B} = \varrho \, \ddot{\boldsymbol{x}},$$

$$\varrho \, \dot{\varepsilon} = \text{tr} \, (\boldsymbol{T} \boldsymbol{D}) + \boldsymbol{\mathfrak{E}} \cdot \boldsymbol{P}^*.$$

By the principle of material frame-indifference, suitably generalized,

$$\hat{\varepsilon} \, (\boldsymbol{Q} \boldsymbol{F}, \, \boldsymbol{Q} \boldsymbol{P}) = \hat{\varepsilon} \, (\boldsymbol{F}, \, \boldsymbol{P}).$$

Choosing $\boldsymbol{Q} = \boldsymbol{R}^T$ yields

$$\hat{\varepsilon} \, (\boldsymbol{F}, \, \boldsymbol{P}) = \hat{\varepsilon} \, (\boldsymbol{U}, \, \boldsymbol{R}^T \boldsymbol{P}),$$

$$= \bar{\varepsilon} \, (\boldsymbol{C}, \, \boldsymbol{\Pi}),$$

where $\boldsymbol{F} = \boldsymbol{R} \boldsymbol{U}$, $\boldsymbol{R}$ being a rotation and $\boldsymbol{U}$ a pure stretch, and

$$\boldsymbol{\Pi} = \frac{\varrho_R}{\varrho} \, \boldsymbol{F}^{-1} \boldsymbol{P}.$$

Hence

$$\dot{\boldsymbol{\Pi}} = \frac{\varrho_R}{\varrho} \, \boldsymbol{F}^{-1} \boldsymbol{P}^*.$$

Substitution into the energy equation yields an identity to be satisfied by the functions $\hat{\varepsilon}$, $\boldsymbol{\mathfrak{g}}$, and $\hat{\boldsymbol{\mathfrak{E}}}$. Much as at the corresponding stage in the thermodynamic theory of visco-elastic materials presented in the last lecture, we can find the most general possible forms of $\boldsymbol{\mathfrak{g}}$ and $\boldsymbol{\mathfrak{E}}$ compatible with this requirement:

$$\boldsymbol{T} = \boldsymbol{\mathfrak{g}} \, (\boldsymbol{F}, \, \boldsymbol{P}) = 2 \varrho \boldsymbol{F} \, \frac{\partial \bar{\varepsilon} \, (\boldsymbol{C}, \, \boldsymbol{\Pi})}{\partial \boldsymbol{C}} \, \boldsymbol{F}^T,$$

$$\boldsymbol{\mathfrak{E}} = \varrho_R (\boldsymbol{F}^{-1})^T \, \frac{\partial \bar{\varepsilon} \, (\boldsymbol{C}, \, \boldsymbol{\Pi})}{\partial \boldsymbol{\Pi}} - \boldsymbol{G} \times \boldsymbol{P}^*,$$

where $\boldsymbol{G}$ is an arbitrary vector. The internal-energy function thus appears as a potential not only for the stress but also for a part of the electromotive intensity.

Using this theory, TOUPIN has found various special solutions illustrating the effect of a strong electromagnetic field on a deformable body. The ordinary theories of piezo-electricity and photo-elasticity are included as special cases, and several new effects are predicted.

As in the work on continuum mechanics without the added complication of the presence of an electromagnetic field, the main tool has been the principle of material frame-indifference. The fact that MAXWELL's equations are not themselves indifferent does not justify an objection against use of the principle for constitutive equations: CAUCHY's equations of motion are not indifferent either, and no one mentions this fact when imposing the principle of material indifference in pure mechanics. It is the marriage of the basic principles of mechanics and electromagnetism that is disquieting. One feels that the axioms upon which particular theories are to be constructed ought to be invariant under the same group.

The generalization of the electromagnetic principles is already known. In the *Classical Field Theories*, TOUPIN followed KOTTLER (1922) in laying down as fundamental, rather than the usual form of MAXWELL's equations, the *Maxwell-Bateman laws* in space time: There exist a vector density σ and an absolute co-variant tensor φ, called the "charge-current field" and the "electromagnetic field", respectively, such that for arbitrary three-dimensional and two-dimensional circuits, respectively,

$$\oint \sigma \cdot d \, \hat{S}_3 = 0,$$

$$\oint \varphi \cdot d \, S_2 = 0.$$

These equations are invariant under arbitrary analytic transformations of the four-dimensional space-time manifold. TOUPIN constructed also an invariant form for the Maxwell-Lorentz aether relations. Mechanics is not so easy to formulate in fully invariant terms. It has been suggested that the real answer is to find a continuum mechanics of general relativity. The principle of material frame-indifference should then be given a local form, corresponding to the relativistic concept of observation. Recently ALDO BRESSAN has proposed a theory of continuous media in generalized relativity, allowing for the simultaneous effects of motion, force, electromagnetism, and energy; he lays down a principle generalizing both material indifference and Lorentz invariance. Under the name "non-sentient response", this principle was proposed independently and motivated in detail by L. BRAGG.

V. The ergodic problem
in classical statistical mechanics

I am often asked why rational mechanics is necessarily continuum mechanics. The answer is that it is not. The cultivators of rational mechanics try to maintain standards of two kinds:

1. The model adopted should mirror nature.

2. The mathematics, besides being rigorous, should match the model in generality.

Both these standards are hard to reach in statistical mechanics. The subject is inherently more difficult than continuum mechanics. In this lecture I shall describe one of the cases, regrettably rare, of success, namely, the treatment of the classical ergodic problem by KHINCHIN and R. M. LEWIS.

BOLTZMANN stated the ergodic hypothesis as follows: "The great irregularity of the thermal motion and the variety of extrinsic forces acting upon bodies make it probable that in virtue of the motion we call heat the atoms of bodies take on all positions and velocities compatible with the equation of energy ...".

It is hard to see how this assertion can have seemed "probable" in 1871; as it stands, it is false, and, today at least, obviously so; we may presume that BOLTZMANN, whose writing was no clearer than his mathematics, really had something more subtle in mind. MAXWELL put forward a modified ergodic hypothesis, more carefully thought out so as to be plausible rather than false, but no rigorous theory has grown from it. The aim of ergodic hypotheses was to justify replacement of time averages along the trajectory of a dynamical system by phase averages, which are averages over initial conditions. In later work, the ergodic hypothesis is discarded, and one considers instead the "ergodic problem", namely, that of finding conditions under which time averages may be calculated by means of phase averages.

Let x and X be points in a measure space, and assume that there is a law of motion T^t such that corresponding to any X a unique x is assigned:

$$x = T^t X, \qquad T^0 = 1,$$

where t is the time. The time average $\hat{f}$ of a function $f(x)$ during the motion is defined as

$$\hat{f} \equiv \hat{f}(X, t_0) \equiv \lim_{t \to \infty} \frac{1}{t - t_0} \int_{t_0}^{t} f(T^\tau X)\, d\tau,$$

if the limit exists. For given X, if the function f is bounded and $f(T^t X)$ is integrable over any interval of t, the mere existence of $\hat{f}(X, t)$ for all t_0 implies that $\hat{f}$ is independent of t_0, i.e., $\hat{f} = \hat{f}(X)$.

The phase average introduced by MAXWELL and called "micro-canonical" by GIBBS is defined as follows:

$$\bar{f} \equiv \bar{f}(u, t) = \lim_{\delta \to 0} \frac{\int_{S_\delta[u]} f(x)\, dm}{m(S_\delta[u])},$$

where $S_\delta[u]$ is a shell of thickness 2δ surrounding the surface $H(x) = u$ on which the energy function $H(x)$ has the constant value u, that is, $S_\delta[u] = \{x : |H(x) - u| \leq \delta\}$, and where $m(S_\delta[u])$ is the measure of the shell $S_\delta[u]$.

If we are to have $\hat{f} = \bar{f}$, for some particular f, it is *necessary* that $\hat{f}$ shall be independent of X for all X on $H = u$, and that $\bar{f}$ shall be independent of t. Now this latter requirement is easy to meet: We need only adjust the assignment of the measure m as time goes on. The microcanonical ensemble is commonly explained by assigning equal *a priori* probability to all initial conditions consistent with the energy u; this becomes a statement with meaning only if we first tell what is meant by probability. There are many ways of assigning mathematically a probability, *i.e.*, a normalized measure, but if expectations according to this probability are to correspond to some physical idea, we demand a physical explanation of it. In the treatises on statistical mechanics this matter is hidden behind unnecessary formalism which goes unchallenged only because we are used to calculations of this kind in Euclidean geometry. At bottom, we are made to assume, without being told what is going on, that the natural assignment of probability is the Lebesgue measure. Beyond a theorem of GIBBS, seldom noticed, which shows that the Lebesgue measure of an ensemble whose motion is governed by the Hamiltonian equations is invariant under canonical transformations, no reason is given.

There is no good reason, and this was not MAXWELL'S approach. *The purpose of statistical mechanics*, for phenomena of equilibrium, *is to calculate time averages*, and the ensemble theory is useful only as a tool enabling us to calculate time averages without knowing how to integrate the equations of motion. The ensemble theory is a mathematical device; we are wasting our time if we try to explain it by itself.

Now we can see at once why the p-q description, the phase space topologically equivalent to Euclidean space, and normalized Lebesgue measure as the underlying probability in finite regions of that space, are introduced. None of these tricks has bearing on mechanics; the laws of mechanics go on governing the system and predicting just the same phenomena, no matter what the description, and no matter what the topology of the phase space used to visualize the results. But a positive solution to the ergodic problem requires, that $\bar{f}$ shall be independent of t, and that $\bar{f} = \hat{f}$, and we seek a phase space, a measure, and a description such as to make these results true.

It is well known that description of a conservative dynamical system in p-q space leads to a steady phase flow, and that Lebesgue measure in this phase space is conserved by the motion. Thus, for the simpler and more commonly studied kinds of system, the usual approach leads effortlessy to a definition of phase average which is independent of time, and the same choice applies to every function f.

Henceforth we shall use Lebesgue measure to define phase averages, but we shall *not* assign any physical meaning to phase averages or to any probability. *Time averages* are what we seek, and we are merely assembling tools to help us find them without knowing any details about the phase motion. We shall assume that T^t is measure-preserving, but no more specific dynamical laws are adopted until later.

Consider the desired result:

$$\hat{f} \text{ a.e.} = \bar{f}.$$

For this to be true, $\hat{f}$ must be generally independent of X, independent, that is, of the particular phase trajectory. Now this cannot hold for all f if there exists an integral, say $y(x) = \text{const}$, such that for some constant a the sets on which $y(x) < a$ and $y(x) \geqq a$ are of positive measures α and $1 - \alpha$. For since $y(x) =$ const. is an integral, no trajectory crosses from one of these sets into the other. Then the time average of the function which is 1 on the former set and 0 elsewhere is 1 for all trajectories in the former set, 0 for all in the latter set, while the phase average of this function is α.

Thus the condition of *metrical transitivity*, viz, that it shall be impossible to decompose the phase space into two invariant parts of positive measure, must hold if $\hat{f} = \bar{f}$ for every f. Note that it is not a matter of knowing the values of any further integrals; it is only the *existence* of integrals that matters.

Almost the only precise information we have concerning the ergodic problem is furnished by the *ergodic theorem* of BIRKHOFF. The assumptions are

1. $x \in \Gamma$, where Γ is the union of a countable collection of sets of finite measure. $m(\Gamma)$ may be finite or infinite.

2. If f is a real-valued measurable function, and if S is a Borel set on the real line, $f^{-1}(S)$ is a measurable set in Γ.

3. The group of transformations T^t of Γ onto itself is measure-preserving for all values of the real parameter t, and the transformation $T^t x$ carries measurable sets in $\{t\} \times \Gamma$ into measurable sets in Γ. Then for any summable function f,

$$\hat{f} \text{ exists a.e.}$$

and is integrable; if $m(\Gamma) < \infty$, then

$$\int_\Gamma \hat{f} \, d\,m = \int_\Gamma f \, d\,m,$$

or

$$\bar{\hat{f}} = \bar{f}:$$

The phase average of the time average equals the phase average itself.

We have seen that metrical transitivity is a condition necessary in order that $\hat{f} = \bar{f}$ for every f. It is also sufficient, for if $G =$ const is any measurable integral, then the set on which $G \geqq a$ is an invariant part, and if $G \not\equiv$ const. a.e., for some a that set will have positive measure, contradicting metrical transitivity. Hence the only integrals are constants a.e. But $\hat{f}$, since it is independent of t, is an integral. Hence $\hat{f} = \text{const.}$ Therefore,

$$\bar{f} = \hat{f} \quad \text{a.e.}$$

For all time averages to be equal to the corresponding phase averages, then, it is both necessary and sufficient that the motion be metrically transitive. This famous result, proved by BIRKHOFF in 1930, has stood for years as a trap, seeming to block any positive outcome to the original questions in ergodic theory, and a verbose and irrelevant physical literature followed. A mathematical theorem cannot be escaped by denying its truth or by forgetting it for vague, intuitive reasons that blur the edges of all rational processes. The way to escape an unpleasant theorem is to prove another one. In fact, the more general theorem of BIRKHOFF affords the key to the solution of the whole problem.

Rather than forcing ourselves to consider only phase averages in calculating time averages, willy nilly (in this case, nilly), let us ask how to calculate time averages in general, without prejudice. The problem of connecting time averages to the nature of the phase motion was first faced, it seems, by R. M. LEWIS in 1960.

If $y(x) = \text{const}$ is an integral for almost all trajectories, *i.e.*, if $y(T^t x) = y(x)$ a.e., we shall call the function y an *invariant* of the motion T^t. Any constant is an invariant, so every motion has invariants. It is clear, moreover, that in the cases of physical interest, when Γ is a space of finite dimension, there will be a finite maximal number of independent measurable invariants. We shall assume that there exists a finite set of invariants,

$$\boldsymbol{y} = \{y_1, y_2, \ldots, y_k\},$$

such that every measurable invariant is functionally dependent upon them; *i.e.*, if z is an invariant, then $z(x) = Z(\boldsymbol{y}(x))$ a.e., where Z is a Borel-measurable function. A set of functions $\boldsymbol{y}$ of this kind will be called a *complete invariant* of the motion.

Now we have seen that $\hat{f}$, if it exists, is an invariant; by the Ergodic Theorem, $\hat{f}$ does exist and is measurable. Hence

$$\hat{f} = F(\boldsymbol{y}(x)) \quad \text{a.e.}$$

Our problem is to find the function F. Given a complete invariant $\boldsymbol{y}$, that is, we are to find the way in which the time average of a particular function may be determined from it.

First, we observe from the Ergodic Theorem itself that if A is any measurable invariant part of $\varGamma$, and if $m(A)<\infty$, we have

$$\int_A \hat{f}\, dm = \int_A f\, dm.$$

Now if B is any Borel set in Euclidean k-space R^k, then $\boldsymbol{y}^{-1}(B)$ is measurable, and since the y_i are invariants, it differs from an invariant part by a set of measure 0. Hence

$$\int_{\boldsymbol{y}^{-1}(B)} F\big(\boldsymbol{y}(x)\big)\, dm = \int_{\boldsymbol{y}^{-1}(B)} \hat{f}\, dm = \int_{\boldsymbol{y}^{-1}(B)} f\, dm.$$

The measure m given in $\varGamma$ induces a measure M in R^k as follows:

$$M(B) = m\big(\boldsymbol{y}^{-1}(B)\big),$$

where B is any Borel set in R^k. For any Borel-measurable F, we have

$$\int_{\boldsymbol{y}^{-1}(B)} F\big(\boldsymbol{y}(x)\big)\, dm = \int_B F(\boldsymbol{y})\, dM.$$

Hence

$$\int_B F(\boldsymbol{y})\, dM = \int_{\boldsymbol{y}^{-1}(B)} f\, dm.$$

We have found the integral of F over an arbitrary Borel set in R^k; now, to obtain F itself, we have only to differentiate in generalized sense.

The result needed is known from the theory of nets. Write

$$I_\delta[\boldsymbol{u}] = \{\boldsymbol{y}: |y_i - u_i| \leq \delta\}.$$

According to a theorem on nets, for almost every u there exists a sequence of δ's such that

$$F(\boldsymbol{u}) = \lim_{\delta \to 0} \frac{\int_{I_\delta[\boldsymbol{u}]} F\, dM}{M(I_\delta[\boldsymbol{u}])} \quad \text{a.e.}$$

This theorem generalizes the familiar idea that the mass density is the ultimate ratio of mass to volume, provided a suitable sequence of shrinking, nested regions be selected.

Writing the result in terms of integrations in Γ, we have

$$F(\boldsymbol{u}) = \lim_{\delta \to 0} \frac{\int_{S_\delta[\boldsymbol{u}]} f \, dm}{m\left(S_\delta[\boldsymbol{u}]\right)} \quad \text{a.e.,}$$

where

$$S_\delta[\boldsymbol{u}] = \boldsymbol{y}^{-1}(I_\delta[\boldsymbol{u}]) = \{x : |y_i(x) - u_i| \le \delta\}.$$

Set $\boldsymbol{u} = \boldsymbol{y}(x)$ and

$$\bar{f}_{\boldsymbol{u}} \equiv \lim_{\delta \to 0} \frac{\int_{S_\delta[\boldsymbol{u}]} f \, dm}{m\left(S_\delta[\boldsymbol{u}]\right)}.$$

Then

$$\hat{f}(\boldsymbol{y}^{-1}(\boldsymbol{u})) = \bar{f}_{\boldsymbol{u}} \quad \text{a.e.}$$

This is *Lewis's theorem*. It states that the value of the time average of f, for almost all trajectories on which the complete invariant $\boldsymbol{y}$ has the value $\boldsymbol{u}$, may be calculated as the phase average over an ensemble in which equal *a priori* probability, in the sense of the basic measure m which the phase motion preserves, has been assigned to every phase in a suitably selected, vanishingly small shell about the integral surfaces $\boldsymbol{y}(x) = \boldsymbol{u}$. If the complete invariant reduces to a single function, say $H(x)$, the results reduce to $\hat{f} = \bar{f}$ a.e.

There is nothing surprising about the more general result. Suppose that a complete invariant $\boldsymbol{y}$ is given by the single function H, the energy function. The motion is then metrically transitive, and LEWIS's theorem reduces to the strong ergodic theorem of BIRKHOFF. The departure of LEWIS from the older line of thinking comes in his giving us a positive rather than negative conclusion when metric transitivity does not hold. We learn what we have to prescribe *in order to get a strong ergodic theorem:* the value of a complete invariant.

Under certain rather elaborate hypotheses, which I shall state in a few minutes, it can be shown that the microcanonical phase average may be approximated by a canonical one:

$$\bar{f}(u) \sim \langle f \rangle_\vartheta = \frac{1}{z(\vartheta)} \int_\Gamma e^{-\vartheta} f \, dm,$$

where ϑ and $z(\vartheta)$ are determined by the conditions

$$\langle 1 \rangle_\vartheta = 1, \quad \langle H \rangle_\vartheta = u.$$

According to Lewis's theorem, the time average of f is not equal to the microcanonical average but rather to a more general phase average, in which the value of a complete invariant is assigned. We shall expect, then, and it is possible to prove, that under appropriate assumptions of uniformity

$$\hat{f}(\boldsymbol{u}) \sim \langle f \rangle_{\boldsymbol{\theta}},$$

where

$$\langle f \rangle_{\boldsymbol{\theta}} = \frac{1}{z(\boldsymbol{\theta})} \int_{\Gamma} e^{-\boldsymbol{\theta} \cdot \boldsymbol{y}(x)} f(x) \, dm,$$

where

$$\langle 1 \rangle_{\boldsymbol{\theta}} = 1, \quad \langle \boldsymbol{y} \rangle_{\boldsymbol{\theta}} = \boldsymbol{u}.$$

Phase averages of this kind, called *polycanonical*, generalize the canonical averages in that a complete invariant $\boldsymbol{y}(x)$ replaces the single energy function $H(x)$, so that in place of ϑ, the scalar index of the canonical distribution, appears the vector $\boldsymbol{\theta}$, a kind of reciprocal temperature vector, which we may call the *coldness* of the distribution. If time averages are expressible approximately as polycanonical phase averages corresponding to a fixed value of the complete invariant, we can now determine the thermostatic structure so implied. Defining a generalized partition function,

$$z = z(\boldsymbol{\theta}, \boldsymbol{\beta}) = \int \exp\left[-\boldsymbol{\theta} \cdot \boldsymbol{y}(x, \boldsymbol{\beta})\right] dm,$$

we find that

$$d \log z = -\frac{1}{z} \int \exp\left[-\boldsymbol{\theta} \cdot \boldsymbol{y}\right] \left\{ \sum_i \left(y_i \, d\theta_i + \sum_k \theta_k \frac{d y_k}{\partial \beta_i} \, d\beta_i \right) \right\} dm.$$

Thus if

$$u_i \equiv \langle y_i \rangle_{\boldsymbol{\theta}}, \qquad X_{ki} \equiv \left\langle \frac{\partial y_k}{\partial \beta_i} \right\rangle_{\boldsymbol{\theta}},$$

we have

$$-d \log z = \boldsymbol{u} \cdot d\boldsymbol{\theta} + \boldsymbol{\theta} \cdot \boldsymbol{X} \cdot d\boldsymbol{\beta}.$$

That is,

$$u_i = -\frac{\partial \log z}{\partial \theta_i}, \qquad \sum_m \theta_m X_{mi} = -\frac{\partial \log z}{\partial \beta_i}.$$

$\boldsymbol{u}$ is the expectation of the complete invariant $\boldsymbol{y}$; alternatively, it is the assigned value of $\boldsymbol{y}$ in the motion of the system and its surroundings. The tensor $\boldsymbol{X}$ is the expectation of the "thermostatic force" corresponding to the change in the value of $\boldsymbol{y}$ resulting when $\boldsymbol{\beta}$ is altered. If we set

$$q = \frac{1}{z} \exp[-\boldsymbol{\theta} \cdot \boldsymbol{y}], \qquad B = -\int q \log q \, dm,$$

then $\log q + \log z + \boldsymbol{\theta} \cdot \boldsymbol{y} = 0$. Multiplying this equation by q and integrating over Γ yields

$$B = \log z + \boldsymbol{\theta} \cdot \boldsymbol{u},$$

showing that the combination $B - \boldsymbol{\theta} \cdot \boldsymbol{u}$ is determined explicitly by the partition function. In particular

$$\boldsymbol{\theta} \cdot d\boldsymbol{u} = dB - d \log z - \boldsymbol{u} \cdot d\boldsymbol{\theta} = dB + \boldsymbol{\theta} \cdot \boldsymbol{X} \cdot d\boldsymbol{\beta}.$$

By definition, B is a function of $\boldsymbol{\theta}$ and $\boldsymbol{\beta}$, but since the relation $\boldsymbol{u} = \langle \boldsymbol{y} \rangle_{\boldsymbol{\theta}}$ may be inverted to give $\boldsymbol{\theta}$ as a function of $\boldsymbol{u}$ and $\boldsymbol{\beta}$, we may regard B as a function of $\boldsymbol{u}$ and $\boldsymbol{\beta}$. Therefore

$$\theta_i = -\frac{\partial B}{\partial u_i}, \qquad \sum_m \theta_m X_{mi} = \frac{\partial B}{\partial \beta_i}.$$

These relations present the form assumed by statistical thermo-statics when a complete invariant, rather than the energy function alone, is given an assigned value. The whole structure is different from the classical one. The temperature, which in the classical theory is interpreted as proportional to the reciprocal of the scalar ϑ, is now replaced by the coldness vector $\boldsymbol{\theta}$. While, perhaps this more general kind of thermostatics is sometimes needed, in most circumstances the usual form seems to suffice. The analysis so far seems to indicate, on the contrary, that if integrals beyond the integral of energy *exist*, the thermostatic properties of the system in weak interaction with its environment differ in consequence. We thus come up short again at an old problem, one of the ugliest which the student of statistical mechanics must face: What can be said about the integrals of a dynamical system? Some of the older works express the idea that however many integrals a system has, generally we shall not know the value of any but the energy, so we should assign equal *a priori* probability to the possible values of the rest — in plainer English, forget about them. Now an idea of this sort, by itself, is just unsound. The results established give us the value of the time average $\hat{f}$; they show unquestionably that *the existence of further time-independent integrals* affects, in general, the value of $\hat{f}$, and whether or not we happen to know about the existence, let alone the values, of these integrals is beside the point; this is, in fact, obvious.

Despite two centuries of study, the integrals of general dynamical systems remain covered with darkness. To save the classical thermostatics, the practical success of which is shown by the wide use to which it has been put, we must find a way out. That is, we must find some mathematical connection between time averages of the functions of physical interest and the corresponding simple canonical averages.

Four different ideas present themselves.

A) Not all measurable functions are of interest in statistical mechanics. By narrowing the class of functions of which ergodic behavior is wished, the analyst may broaden the class of phase motions for which an ergodic theorem holds. Rather than "ergodic motion" we should seek the "ergodic functions" in a given motion.

B) It is not necessary that $\hat{f} - \langle f \rangle = 0$ exactly. All that is needed is that $\hat{f} - \langle f \rangle$ be small in a precise way.

C) The integral of energy is different from such other integrals as there may be. First, knowledge of the function H specifies the dynamical system entirely, while knowledge of other invariants, such as the linear momentum and the angular momentum, does not. In other words, the form of the energy integral *must* be known; whether or not other integrals exist follows, in principle (though, unfortunately, not in practice), from the form of the energy integral.

D) While the developments given so far are valid equally for all kinds of systems, large or small, the physical systems to which statistical thermostatics is considered applicable consist of very many particles.

KHINCHIN'S solution of the ergodic problem makes use of all four of these ideas. Since his result is now well known, I shall not present more than a sketch of the mathematics, but I wish to make clear the physical thinking behind it. Of course, the fourth idea, bringing out the *asymptotic character* of all the non-trivial results of statistical mechanics, is the real key, but the argument is easier to follow if we state the assumptions separately from their asymptotic interpretation. The asymptotic nature of the problem will be left for the end.

To narrow the class of functions, we recall that the essential step in the proof of the strong ergodic theorem was to show that $\hat{f}$ has the same value for almost all trajectories. Instead of requiring this constancy for all f, we consider those particular f for which it holds. By the Birkhoff theorem, $\overline{\overline{f}}=\overline{f}$; hence $\overline{f}=\hat{f}$ a.e. That is, if $\hat{f}=\text{const}$, then f is an ergodic function.

We are now in a position to relax the requirement that $\hat{f}=\text{const}$ by supposing that $\hat{f}-\text{const.}$ is "small"; it will then follow that $\hat{f}-\overline{f}$ is the mean value of a small function and hence small. This is the type of result we can reasonably expect to establish. To do so, we must give a precise definition of "small". There may be various possible ways, but the apparatus of ergodic theory makes it easiest to think of smallness in terms of the measure in phase space. Let P be a probability, that is, a measure such that $P(\Gamma)=1$; the earlier hypotheses regarding the space Γ and the motion T^t are retained. Then what we wish is a statement that

$$P(|\hat{f}-\overline{f}|\leq\varepsilon)\leq\delta$$

for suitably related ε and δ, and where $\overline{f}=\int f\,dP$.

As noticed by KURTH, it follows from BIRKHOFF's theorem that a function which is nearly constant is, in general, still more nearly ergodic. For by the Schwarz inequality we have, for any summable h,

$$(\hat{h})^2\leq\widehat{h^2}.$$

Taking phase averages and then applying the Birkhoff theorem yields

$$\overline{(\hat{h})^2}\leq\overline{\widehat{h^2}}=\overline{h^2}.$$

Setting $h=f-\overline{f}$, we see that

$$\overline{(\hat{f}-\overline{f})^2}\leq\overline{(f-\overline{f})^2}:$$

The phase dispersion of the time average $\hat{f}$ can never exceed the phase dispersion of the function f, or, in other words, time averaging on most trajectories smooths out deviations from phase averages.

To make use of this result, we apply the inequality of Bienaymé and Chebichev, which estimates the probability of phase deviation in terms of phase dispersion, *viz*

$$P(|h-\overline{h}| \geq \alpha) \leq \frac{1}{\alpha^2} \overline{(h-\overline{h})^2}.$$

Set $h = \hat{f}$. Since, by Birkhoff's theorem, $\overline{h} = \hat{f}$, we see at once that

$$P(|\hat{f}-\overline{f}| \geq \alpha) \leq \frac{1}{\alpha^2} \overline{(f-\overline{f})^2}.$$

Since α is arbitrary, we may set $\alpha = K \sqrt[4]{\overline{(f-\overline{f})^2}}$. Then

$$P(|\hat{f}-\overline{f}| \geq K \sqrt[4]{\overline{(f-\overline{f})^2}}) \leq \frac{1}{K^2} \sqrt{\overline{(f-\overline{f})^2}}.$$

In other words, a nearly constant function is very probably a nearly ergodic function.

Recall that we agreed that it is time averages we wish to evaluate. In the ergodic theory of Lewis, a particular measure for phase averages presented itself as a natural consequence of the mathematics, and we met no necessity of introducing ensembles as such. Now the situation is changed. The phase averages in the last inequality refer to any invariant probability: No particular one is suggested. To make use of the result, we must assign a probability such that

1. The phase dispersion is small for those f which are of physical interest.

2. A set of trajectories of small probability may be considered unlikely to occur. *I.e.*, the probability is a physically natural one.

Unfortunately, the second requirement forces us to enter some compromise with the ensemble theory. It is still time averages that we wish, but we must agree on the kind of trajectories we shall regard as unimportant enough to neglect. Granted that a dynamical system has the complete invariant $\boldsymbol{y} = (y_1, y_2, \ldots y_k)$, we recognize formally in the theory that in general we cannot determine this invariant. We divide the complete invariant into two parts, the *controllable integral* $H = (y_1, y_2, \ldots, y_l)$ and the *residual invariant* $(y_{l+1}, \ldots, y_k)$. The controllable integral consists in the maximal number of functionally independent, measurable invariants whose form is known to us and *whose value may be assigned* as part of the statement of the physical problem. Usually the controllable integral will consist of the energy function alone, but if we are able to assign the values of other integrals, such as those of linear momentum and of moment of momentum, the theory must be broad enough to let us do so.

We should like to be able to say that the value of the residual invariant does not matter. From LEWIS' results, however, we know that *this is not true*. To escape the effect of possible further unknown integrals in calculating the time averages of certain particular functions by means of KHINCHIN'S approach, we must accept a class of trajectories as unimportant — *viz*, the class for which the final inequality gives no information. The only way I see of doing this is to regard the value of the residual integral as itself a random variable insofar as exceptional trajectories are concerned, and to assign to it an equal *a priori* probability in the sense of the measure that results from the ergodic theory in the case when the controllable integral is a complete invariant. In other words, we agree to accept the *polymicrocanonical probability* based on the controllable integral alone, not for an underlying ensemble theory but only as a criterion for trajectories to be considered of no importance. We might call this the *null ensemble hypothesis: In calculating time averages, we are content to get a result valid except on a set of small polymicrocanonical probability defined in terms of the controllable integral alone.* I believe that this is the minimum compromise we can make with the ensemble approach. I believe also that it comes close, granted a precise mathematical expression, to the ergodic hypothesis as stated by MAXWELL. It carries with it a clearer view of what the statistical aspect of statistical mechanics really is, namely, the assertion of *circumstances which may be neglected*. Let us put the matter in terms of an "experiment" in which we can fix H but no more; if the "apparatus", unknown to us, fixes the value of a single further integral, then all results from this apparatus will refer to a set of measure 0 in the probability we have assigned, and the conclusions from KHINCHIN'S approach will have no physical relevance at all. This should not be too shocking. Sometimes a theory is applicable, sometimes it is not, and, once the theorems are proved, it is up to the applier to use his judgment in applying it. Part of the trouble with statistical mechanics lies in its proponents' annoying and unjustified claims for its overwhelming universality. Like most of the fundamental assumptions of physics, the null ensemble hypothesis is not conceivably open to direct experimental test. The ergodic theory shows that such is the price we must pay for getting, for a particular f, an ergodic theorem which is simpler than the laws of dynamics give us any right to expect in general.

To complete the proof of the ergodic theorem conjectured, we have to show now that the microcanonical phase dispersion of suitable functions f is really small. While a result of the same kind may be expected for an arbitrary controllable integral $\boldsymbol{H} = (y_1, \ldots, y_l)$, the details have been worked out only for the case when $\boldsymbol{H}$ reduces to the energy function alone, so only this case will be considered further.

While we have faced and emasculated the statistical aspect of the ergodic problem, we have left the asymptotic aspect until now. We attempt to put in mathematical form the idea that thermodynamics is appropriate to description of the behavior, in the sense of time averages, of systems having very many degrees of freedom. Since there is not known a single result from analytical dynamics indicating any difference of kind in the behavior of a system with 10^{10} degrees of freedom from one with 10, we take refuge in asymptotics. We construct a sequence of ever larger dynamical systems, and for each of these we calculate the microcanonical phase dispersion of a class of functions f. If we can find conditions of uniformity such that for the f under consideration, $\overline{(f-\bar{f})^2} \to 0$ in this sequence, it will follow that

$$P(|\hat{f} - \bar{f}| > \varepsilon) \to 0,$$

where ε is as small as we like. That is, any inaccuracy in using the microcanical phase average $\bar{f}$ to evaluate the time average $\hat{f}$ becomes more and more unlikely, the larger is the system — more unlikely in the sense of the probability used to define the phase average.

KHINCHIN used the same apparatus to evaluate the dispersion as he had used previously to prove that, in certain circumstances, $\bar{f}$ may be approximated by an appropriate canonical average, $\langle f \rangle_\vartheta$. In a word, he showed that the ergodic property is a refinement of BOLTZMANN'S law.

First, it is assumed that the energy function is of the form

$$H(x) = \sum_{\nu=1}^{n} H_\nu(x_\nu),$$

so that the system consists of n independent parts with phase spaces Γ_ν and $\Gamma = \prod_{\nu=1}^{n} \Gamma_\nu$, $x_\nu \in \Gamma_\nu$. In the physical literature a Hamiltonian of this form is often alleged to represent sufficiently a system of n "weakly interacting" constituents. It is assumed that H_ν has a continuous and non-vanishing gradient, and that each phase x_ν lies upon one and only one surface $H_\nu(x_\nu) = u_\nu \geq 0$, and that the region in which $H_\nu(x_\nu) < u_\nu$, $u_\nu > 0$, is finite and simply connected. This requirement, that the enregy surfaces $H_\nu = u_\nu$ shall sweep out the phase space Γ_ν like a family of spheres centered upon the origin, is satisfied by the typical Hamiltonian functions used in mechanics. In consequence of it, phases for which $|x_\nu|$ is large are phases of large energy. If we set

$$\omega_\nu(u_\nu) \equiv \int\limits_{h_\nu = u_\nu} \frac{dS_\nu}{\text{grad } H_\nu},$$

if follows that for a function $f_\nu(x_\nu)$ the microcanonical expectation when $H = u$ is given by

$$\bar{f}_\nu = \frac{1}{\Omega(u)} \int\limits_{\Gamma_\nu} f_\nu(x_\nu) \Omega_\nu(k - H_\nu(x_\nu))\, dm_\nu,$$

where m_ν is Lebesgue measure in Γ_ν, where Ω_ν is the convolution of all the ω_μ except ω_ν, and where Ω is the convolution of all the ω_ν. As the last requirement to be imposed on the dynamical system, we assume that $\omega_\nu(u)$ is a differentiable function of the real variable u and that there exists an L_ν such that

$$\omega_\nu(u) = O(u^{L_\nu}) \quad \text{as} \quad u \to \infty.$$

If we set

$$g_\nu^{(\alpha)}(u) \equiv \begin{cases} \dfrac{1}{s_\nu(\alpha)} \exp[-\alpha u]\, \omega_\nu(u) & \text{if} \quad u \geq 0, \\ 0 & \text{if} \quad u \leq 0, \end{cases}$$

$$s_\nu(\alpha) \equiv \int\limits_{0}^{\infty} \exp[-\alpha \xi]\, \omega_\nu(\xi)\, d\xi,$$

for any fixed positive α, then it is easy to see that the $g_\nu^{(\alpha)}$ are densities of independent random variables and are differentiable functions having finite moments of all orders. The corresponding functions G_ν and G are densities for the sum of $n-1$ and n independent random variables. The central limit theorem may be applied to evaluate these densities asymptotically as $n \to \infty$; asymptotic forms for Ω_ν and Ω follow at once, and an asymptotic form for $\bar{f}_\nu$ results.

In addition to the hypotheses already made regarding the system itself, two further kinds of hypotheses are needed in order to get a manageable result.

Hypothesis regarding f_ν: There exists a constant K such that

$$f_\nu(x_\nu) = O\left([H_\nu(x_\nu)]^{K_\nu}\right)$$

as $x_\nu \to \infty$. That is, we agree to be content with averaging only those functions which do not increase too much faster than the energy in Γ_ν'.

Hypotheses of uniformity in the comparison process:

1. As $n \to \infty$, only finitely many of the H_ν are different. That is, we compare ever larger systems of like composition.

2a. (Maxwell-Boltzmann process):

$$\sum_{\nu=1}^{n} H_\nu(x_\nu) = u = \text{const.}$$

2b. (Gibbs process):

$$\sum_{\nu=1}^{n} H_\nu(x_\nu) = u = u_0 n, \qquad u_0 = \text{const,}$$

$$\lim_{u \to \infty} \frac{1}{n} \sum_{\nu=1}^{n} \frac{s_\nu'(\alpha)}{s_\nu(\alpha)} \text{ exists.}$$

Conditions 2a and 2b are alternatives. In the former, the total energy is held constant as $n \to \infty$, so that the arithmetic mean energy u/n of the system in Γ_ν' grows smaller and smaller. We are to think of a sequence of ever more numerous dynamical systems serving as the model for behavior of one physical system of given total energy. In the Gibbs process, the arithmetic mean energy u/n of the system in Γ_ν' is held constant. We are now to think of the sequence of ever more numerous dynamical systems as provid-

ing a model for a particular physical system plus its environment. As the environment is taken large and larger, so is the total
energy of the system and environment together.

In both cases, a result having the form of BOLTZMANN'S law
has been shown to follow:

$$\bar{f}_\nu = \langle f_\nu \rangle + O\left(\frac{1}{n}\right).$$

This formula is the key to statistical thermostatics.

Most functions of interest in statistical mechanics are sum
functions, having the form

$$f(x) = \frac{1}{n} \sum_{\nu=1}^{n} f_\nu(x_\nu),$$

where $f_\mu = f_\nu$ if $H_\mu = H_\nu$. Then

$$\bar{f} = \frac{1}{n} \sum_{\nu=1}^{n} \langle f_\nu \rangle + O\left(\frac{1}{n}\right).$$

On the final examination for a course of mine in 1955, Mr. MOR-
GENSTERN showed me how to replace, by a few simple lines,
KHINCHIN's chapter estimating the dispersion of such a function:
First

$$\bar{f}^2 = \frac{1}{n^2} \sum_{\nu=1}^{n} \langle f_\nu \rangle^2 + \frac{1}{n^2} \sum_{\mu \neq \nu} \langle f_\nu \rangle \langle f_\mu \rangle + O\left(\frac{1}{n}\right) = \frac{1}{n^2} \sum_{\mu \neq \nu} \langle f_\nu f_\mu \rangle + O\left(\frac{1}{n}\right),$$

since it is a property of the canonical distribution that $\langle f_\nu \rangle \langle f_\mu \rangle =$
$\langle f_\nu f_\mu \rangle$. Now regard the systems in Γ_μ and in Γ_ν as a single system,
and consider for it the function $f_\mu(x_\mu) f_\nu(x_\nu)$. By BOLTZMANN's law,

$$\overline{f_\mu f_\nu} = \langle f_\mu f_\nu \rangle + O\left(\frac{1}{n}\right).$$

But

$$\overline{(f-\bar{f})^2} = \overline{f^2} - \bar{f}^2 = \frac{1}{n^2} \sum_{\nu=1}^{n} \overline{f_\nu^2} + \frac{1}{n^2} \sum_{\mu \neq \nu} \overline{f_\mu f_\nu} - \bar{f}^2.$$

Substituting the above estimates in this formula yields

$$\overline{(f-\bar{f})^2} = O\left(\frac{1}{n}\right),$$

which was to be proved.

That is, in comparison sequences of the types defined, it turns
out that a sum function is more and more nearly constant as n

increases. By KHINCHIN's estimate, it follows that the trajectories on which f fails to be ergodic grow more and more unlikely in the sense of the microcanonical probability. If we agree, then, to be content with results valid except on trajectories of microcanonical probability zero, we can summarize the results, subject to the hypotheses stated regarding the sum function f and the comparison sequence selected:

$$\hat{f} \approx \bar{f} \approx \langle f \rangle_\theta,$$

and statistical thermostatics follows in its classical form. If, however, we have additional information about the system, such as the exact value of an additional invariant, the theorem loses its relevance, since then, generally, only trajectories of microcanonical probability zero will interest us.

I should like to be able to say that statistics is unnecessary, that the name "statistical mechanics" is a misnomer for "asymptotic mechanics", as far as equilibrium is concerned. This is almost true, but not quite so.

VI. Method and taste in natural philosophy

I am frequently asked to define rational mechanics or natural philosophy. The questioner usually wishes particularly to know what is excluded. Is everything else "irrational" or "unnatural"? In fact, the revival of both these old terms began as a measure of defense, not seeking to exclude any particular area but rather to find a name covering something excluded by everyone else.

In 1946 I was employed as an adjunct to a large captured wind tunnel, where my interest was directed to rarefied gases. The gaseous chief, more dense than rare, was unable to place my efforts in any pigeonhole. With the professional certainty of a former assistant professor of physics at a minor degree mill, he knew that what I did was not physics. While his senior aerodynamicist assured him it was the purest of pure mathematics, an aging refugee estimator of "eigenvalues" begged to be relieved from evaluating my work on the ground that he himself was a mathematician. Indeed, several mathematician friends told me that any paper in which the words "stress" or "vorticity" appeared was clearly engineering or physics. Those few engineers I ran across were too polite to damn my area of study completely, but they did say that engineering application for it lay at least 200 years in the future.

It is pleasant to afford the luxury of being an eccentric, but no one likes to be nothing. I sought a name that would reflect mathematical approach to problems of the motion of masses and found it, naturally enough, in the designation NEWTON used for his own work on this area. The term "rational mechanics", coined by the ancients, was discarded in English only when science fell into professions. Far from seeking to exclude any area, use of the term "rational mechanics" indicates an interest *broader* than any of today's specialities, but no less precise. As time went on, the methods and views developed in rational mechanics proved useful in thermodynamics, electromagnetism, and relativity, so a still broader term of equal age and standing was sought and found in "natural philosophy", which includes all the mathematical sciences of natural phenomena.

A chemist who lectures before you is not expected to define chemistry, nor is his professional label interpreted as a boast that he knows everything in the entire domain so titled. In this series of lectures I have given examples of the new work in natural philosophy, trusting that they will serve in lieu of a definition, and that my own ignorance of many parts of science needs no further emphasis. At the same time, it is surely clear that "natural philosophy" is neither an alternative term for "general science" nor a mere collective binding for the loose sheaves of all the volumes in the library. Natural philosophy must not be confused either with the slender "interdisciplinary" bridges of uncomprehending mutual cordiality now popularly cast across the rotten foundation left behind in the headlong rush of physics, chemistry, and biology toward the most hazardous frontiers. Attempting neither to throw all scientific activity into one miscellany bag nor to prohibit any specialist who so desires from dancing on the point of his favorite scientific pin, modern natural philosophy seeks to reset the foundation itself, joining to it the distant arms of a sound structure.

As is clear from the foregoing lectures, unity is reached through mathematics. LEIBNIZ wrote in the first paper on the mathematical theory of elasticity, published in 1685, "whence, after consideration of these few things, this whole matter is reduced to pure mathematics, which is the one thing to be desired in mechanics and physics." While many theorists will accept this statement thoughtlessly as a program, it is more than a cliché. The new work in natural philosophy reinterprets it in two ways.

First, the quality of mathematical thought put into natural science has fallen over the centuries. HUYGENS, NEWTON, LEIBNIZ, JAMES BERNOULLI, EULER, and CAUCHY were the greatest mathematicians of their times, creative equally in "pure" mathematics and in applications, and it is they who gave the most to mechanics as we know it today. POISSON, STOKES, KELVIN, HELMHOLTZ, and MAXWELL did some fine original work in mathematics and knew all there was then to know about it. RAYLEIGH showed the bravura of a great mathematician without the discipline and taste which, up to his time, had always accompanied it[1]. PRANDTL was no mathematician at all. The survivors and epigones of the Rayleigh-Prandtl schools, some of whom, bending beneath their honors, are still alive today, not only have no idea what the mathematics of our time is about but also know in absolute quantity less than did STOKES and MAXWELL. These "leaders" interpret mathematics as a dead language and regard the problem of expressing nature in mathematical terms like that of writing a poem in classical Latin, where the subject must be chosen or wrested so as to fit an ancient vocabulary. Nature is expected to adjust its comportment to the mathematics learnt in their school days by the members of some senior international committee for mutual congratulation and world-wide sight-seeing. They often boast that, like Antaeus, they keep their feet on the ground, stoutly refusing to fly off into the rarefied azure of "pure" mathematics. Their performance, however, gives no sign of Antaean stature and suggests rather the dance of an earthworm, who never lifts his head out of the mud. Their highest flights of imagination lead them to guess what the answer in some knotty special case should be and then somehow to truncate and torture the equations of a time-worn

---

[1] In a general apology for mathematics in 1741, EULER contended that " ... the usefulness of mathematics, commonly allowed to its elementary parts, not only does not stop in higher mathematics but in fact is so much the greater, the further that science is developed." In 1867 KELVIN & TAIT, speaking especially of the work of STOKES in 1842—1846, claimed conversely that the best pure mathematics is suggested by application: "... astonishing theorems of pure mathematics, such as rarely fall to the lot of those mathematicians who confine themselves to pure analysis or geometry instead of allowing themselves to be led into the rich and beautiful fields of mathematical truth which lie in the way of physical research." For RAYLEIGH, pure mathematics was already something to sneer at. "For, however important it may be to maintain a uniformly high standard in pure mathematics, the physicist may occasionally do well to rest content with arguments which are fairly satisfactory and conclusive from his point of view. To his mind, exercised in a different order of ideas, the more severe procedure of the pure mathematician may appear not more but less demonstrative." While the turn of "pure" mathematics at the date of writing, 1894, the heyday of uniform convergence and ε-δ estimates, easily arouses our sympathy for the feeling expressed, over the following half century it was physical science, not mathematics, that suffered from this facile escape from facing the facts.

theory until they can squeeze from it some rough agreement with their triumphant "physical intuition". Typically, "pure" mathematics is dismissed as being "useless abstraction" or "chess playing". Calculus was dismissed in this way in the eighteenth century, and partial differential equations in the nineteenth, but only by "practical" men, not by leaders of science. On the other hand, while the great mathematicians of the past often turned their attention directly to nature and were always pleased if physical applications of their results were found, the typical "pure" mathematician of today boasts his entire ignorance of natural science and his contempt for application to it. In some cases results which might be useful to others if understood he phrases as if by intention in such a way as to bar all but initiates. Nevertheless, it is the work of the "pure" mathematicians, even if they themselves avow the fact with reluctance or even turn their backs upon it, that has brought us the main advances of our own day in applications to natural philosophy or at least put in our hands the tools to effect them.

The first aim of modern natural philosophy is to describe and study natural phenomena by the most fit mathematical concepts. The most fit need not be the most modern, but they may be; indeed, since we are all, whatever our labels, actors on a common stage, they are likely to be. In paraphrase of the program of KELVIN & TAIT in their *Treatise on Natural Philosophy*, we neither seek nor avoid the most abstract mathematics. To use modern mathematics when it is appropriate, one must first acquire it and then see where it applies. Neither to learn nor to use is easy, and I cannot name any one person who comes near to the ideal of the modern natural philosopher. However, the tide has turned, clearly turned, and it is easy to point to at least half a dozen men, none yet over forty, who today are bringing to bear endowments in mathematics such as have not been seen in research on mechanics for half a century. While the specialists in "applied mathematics" or "applied mechanics" continue to tint blackboards and pages grey with routine calculations of extreme "practical importance", described in terms which would have been rejected as vague and loose by CAUCHY, their young students, politely refraining from comment, begin to learn and apply the mathematics of today's mathematicians as a matter of course.

When a student I heard lectures by an elderly faker who enjoyed, and perhaps still posthumously enjoys, a big name as a historian and philosopher of mathematics. He often said that "pure" mathematics had developed because in the nineteenth century the mathematicians had solved physical problems faster than physicists could pose them, so they turned to a new source of study within themselves. With no wish to estimate anyone's motives, I remark that the fact asserted is more than ordinarily untrue. Looking back at the matter of the foregoing lectures, I see rather a parallel to the physical problems suggested by the seventeenth century, which had to wait 100 to 150 years for the theories of partial differential equations and differential geometry to be created before an adequate mathematical treatment could be given. The phenomena represented in the new natural philosophy, with a few spectacular exceptions, were more or less known a century ago. What was lacking was *mathematics* sufficient to describe them adequately and discuss them with profit. It is now an established fact that mathematics and physics can go their separate ways. While professionalized division of these sciences clearly fosters volume of publication, its influence on quality may not be so happy. Certainly the mathematics used in modern natural philosophy might have been developed more quickly, had mathematicians paid attention to natural problems, and certainly the work I have described in these lectures would be less inaccessible to physicists, had the teaching of physics put some value on keeping abreast of the mathematics of one's own time. In attempting to draw mathematics and physics together again, the natural philosopher of today expects and receives no help from either of the organized professions, but the success already achieved, no more than a start, shows the times are ready for what is being done.

The second difference in method lies deeper. Most physical scientists regard mathematical treatment as belonging only to a later stage in the development of theory. While they may appreciate the need for mathematics, perhaps even quite fancy "pure" mathematics, in solving specific problems, they regard the basic principles of the theory as originating in intuition, experiment, or higher authority. For example, a mathematical physicist may spend his whole life on diffraction theory, in the course of which he finds it necessary to learn abstract measure theory, functional analysis, Banach algabra, and the theory of distributions in all precision, yet when he comes to derive the wave equation for his students, he is likely to use reasoning that would have been discarded as unrigorous by EULER. For such a man, mathematics is a device for calculating examples, perhaps very hard ones, but it plays no part in discovering the physical theory. Mathematicians, on the other hand, often regard all of physics as a kind of divine revelation or trickery, where mathematical morals are irrelevant, so that if they enter this red-light district at all, it is only to get what they want as cheaply as possible before returning to the respectability of problems purely mathematical in the older sense: analysis, probability, differential geometry, *etc.* In modern natural philosophy, *the physical concepts themselves* are made mathematical at the outset, and mathematics is used to *formulate* theories.

The usual dividing line for mathematics stops just after geometry, cutting off mechanics. "Space", "point", "line", "figure", "incident", "parallel", "volume", and "area" are mathematical terms, having properties embodied in specific, unequivocal axioms, with no mention of experiment or observation. "Mass" and "force" belong to physics. No mathematical axioms are laid down for them, and much is said about how to measure and interpret them. Specific theories of the action of forces on masses are set up by drawing pictures or repeating rituals, and the mathematics does not come until afterward. It is not a question of rigor: When finally the mathematics gets started, it may be entirely precise, but setting up the theory is an extra-mathematical operation. Sometimes it is said that while geometry is a part of mathematics, mechanics and the rest of physics are experimental sciences.

Although this statement may describe current practice, it is not a divine order. There is an experimental geometry, although few people cultivate it now. At one time, it was doubtless quite a problem to calculate areas and volumes, and for some odd shapes experimental methods are used even today. However important this practical side may be, in leafing through the Johns Hopkins catalogue I do not find a course in experimental geometry listed among the several dozen offerings by the departments of mathematics, physics, mechanics, *etc.*, though I admit my patience gave out before I reached sociology and sanitary engineering. Rather, mathematical geometry has been so widely accepted that we are led to believe we understand all of geometry when we have mastered the mathematics of it. No laboratories are provided where the student may test his knowledge of areas by cutting out and weighing sheets of tin with curved boundaries. Indeed, even the most practical of practical men is likely to have a rather inverted picture of geometry. When the carpenter uses a little more wood than planned, he does not say, "The theory gave the wrong answer, since it neglects some of the real physics," but "The boards weren't square." Rather than try to fit theory to practice, he reproaches application for failing to reach the precision of theory.

Rational mechanics is an extension of geometry, with similar aims, methods, and results. Both are sciences of experience. Neither is experimental. Experimenters and appliers, be they physicists, engineers, or practical men, make good and successful use of both sciences. It has been so many centuries since geometry was distilled from its experiential basis that no one doubts its exactness or asks how or when to apply it. I have never heard of a geometry teacher's being troubled by an engineer who says, "Look at this beam and tell me whether its cross-section is a triangle or a circle!" The practical man has his own methods of setting theory into correspondence with experience; indeed, such correspondence furnishes his specific business; he does not expect geometry itself, or the geometer, to tell him *when* to use $A = d^2$, instead of $A = \pi r^2$ to find the area of a particular table top. Neither does he discourage the geometer with an accusation like those thrown commonly today upon an advocate of continuum mechanics: "Your theory represents a triangle as having smooth, continuous edges, and as being of no thickness, but any real triangle is made up of molecules, its edges are all ragged with holes and gaps, and it has at best some approximate or average thickness. How can your geometry be of any use for physics if it neglects these fundamental facts about the structure of matter?" The practical man neither despises geometry as useless nor worships it as an imperial pill to purge all his troubles and render his function null.

Why is it that the physicist who is content to learn geometry in all precision reverses his stand the moment the concepts of mass and force are added to space and distance, crying that when "physics" enters the picture, rigor must be kicked out by "intuition", and wholly different methods and criteria are in order? He is failing to see that while physics through the centuries has taken over many domains formerly thought parts of philosophy, as the second stage mathematics takes over domains formerly thought parts of physics. The intuitive physicists are the scouts of natural philosophy. While some teaching today suggests that stapling together the loose dispatches from the frontier makes the best of maps, students of a later day learn geography, not from the exciting sketches sent back by courageous pioneers of no-man's land, but from a careful and precise atlas, in which early rumors are sometimes denied as false, sometimes drastically corrected. (Lest the comparison seem not too flattering to the mathematicians, let me add that often the report of one trained surveyor outlasts the braying of a hundred heroes.) Geometry was consolidated as mathematics two thousand years ago; conquest of mechanics for mathematics began three hundred years ago and is now in progress, against resistance. Indeed, often physical scientists, instead of learning and using modern mechanics as they do geometry, with gratitude for what it is, decry it for what it is not, either reproaching the theorist for "neglecting the physics" or suggesting it would be better if he "stuck to mathematics", by which they mean infinitesimal extrapolations of the mathematics they themselves half learned in school.

Whenever nowadays a clean theory is proposed, some physical scientist rushes up to cry about what is left out. Everyone agrees in wishing to understand and explain more. When the physical scientist tells us, as with a faithful monotony he does, that he knows already all there is to know "in principle" about "classical" mechanics, we see a choice between explaining everything badly or a few things well. The experimenter would scorn a suggestion that his grandfather's experimental apparatus suffices for today's measurements, but he seems to be content with his great-grand-father's ideas of mechanics. The applied scientist who regards the new natural philosophy as an enemy union calling for a closed shop mistakes its nature and function. Every day, almost every minute, he uses the simple ideas and results of "classical" physics as tools for research or application. The new natural philosophy attempts to put into his hands methods and concepts both sharper and more flexible, methods and concepts which stand to the old "classical" physics of the textbooks as modern power tools do to hand saws and planes, methods and concepts which can do all that the old did and much more, but which the physical scientist must first *learn* and then use.

Rational mechanics is mathematics, just as geometry is mathematics. The generality and abstraction now raising howls of scorn or pain from the old guard, some members of which, I must add, are arithmetically young, is only common and natural in the course of mathematics. To construct the real number system requires a more abstract level of thought than do operations with rational numbers; all experimental data is reported in terms of rational numbers alone; yet for the past two hundred years even grammar school children have been taught to operate with real numbers, because it is simpler to do so, and the disadvantages of an arithmetic excluding irrationals are too obvious to need detailing. It is the process of generality and abstraction that reduced analytical mechanics, some two hundred years ago, to a form which can be taught to university students today in one semester. There is no need to promote the new and abstract approach to materials. Its advantages are so clear as to want no propaganda, and its victory is certain. The resistance of the old guard is that which has greeted every advance in mathematical science and always dies away with one generation. Already I have noticed at this university that our graduate students take to the abstract and general forms at once, and they have only a patronizing smile for me if I try to explain what seemed to be the difficulties only ten years ago. When I was a student, the older professors of physics spoke of the same experience with quantum mechanics.

It is often remarked that the mathematical community ignores the recent work on natural philosophy, while some engineers welcome and apply it. Apathy of mathematicians toward a new area of mathematics need not be fatal to it, for in the past hundred years there has been no time at which the most original work being done found favor with the profession as a whole. Transfinite numbers, tensor analysis, set theory, and abstract topology all met at first outright hostility from the profession that thrives upon them today. To those who have been active in creating the new natural philosophy, the friendliness of the engineers is a gratification, their interest a flattering stimulus, but the methods and objectives of research remain unchanged. A painter is pleased if his canvas serves as inspiration for a poet or is bought for a pretty penny to grace the home of an oil magnate, but he does not begin himself to write verse or drill for oil. Unfortunately the engineers seem to be more like poets than magnates when it comes to expressing their appreciation of the efforts of natural philosophers.

The ingratitude of applied scientists for the fundamental work they themselves appropriate, often after a period of rejecting it as abstract and "useless", and later turn to profit in their more lucrative professions, sometimes grows out of failure to understand what it really is they have digested. Most natural scientists have a strong bias toward experiment; all of us have been subject to footnotes, books, or even courses in the history of science that distort the growth of scientific thought in the past as wickedly as a communist text distorts political history.

The hard facts of classical mechanics taught to undergraduates today are, in their present forms, creations of JAMES and JOHN BERNOULLI, EULER, LAGRANGE, and CAUCHY, men who never touched a piece of apparatus; their only researches that have been discarded and forgotten are those where they tried to fit theory to experimental data. They did not disregard experiment; the parts of their work that are immortal lie in domains where experience, experimental or more common, was at hand, already partly understood through various special theories, and they abstracted and organized it and them. To warn scientists today not to disregard experiment is like preaching against atheism in church or communism among congressmen. It is cheap rabble-rousing. The danger is all the other way. Such a mass of experimental data on everything pours out of organized research that the young theorist needs some insulation against its disrupting, disorganizing effect. POINCARÉ said, "The scientist must order; science is made out of facts as a house is made out of stones, but an accumulation of facts is no more a science than a heap of stones, a house." Today the houses are buried under an avalanche of rock splinters, and what is called theory is often no more than the trace of some moving fissure on the engulfing wave of rubble. Even in earlier times there are examples. STOKES derived from his theory of fluid friction the formula for the discharge from a circular pipe. Today this classic formula is called the "Hagen-

Poiseuille law" because STOKES, after comparing it with measured data and finding it did not fit, withheld publication. The data he had seem to have concerned turbulent flow, and while some experiments that confirm his mathematical discovery had been performed, he did not know of them.

I am not suggesting that no experiments should be done; since experiment is now a recognized profession, I am not even saying that experimenters are doing too many experiments. I am merely warning the theorist to leave experiment to the experimenters. There is no danger that the theorist will dream up models of a non-existent universe. Abundant experience is at his disposal. The role of theory is not to adjust fudge factors to data, but, like geometry, to draw much from little. Herein lies its economy, its beauty, and, ultimately, its use. EULER in deriving the laws of motion of a rigid body remarked that they would be just as valuable if there were no rigid bodies in the physical world. Once the general ideas of mechanics are understood, certain particular materials, such as the rigid body, the perfect fluid of EULER, and the simple fluid of NOLL, present themselves as naturally as do triangles and circles in geometry, and it is a mathematical necessity to study them. Nature does not seem full of circles and triangles to the ungeometrical; rather, mastery of the theory of triangles and circles, and later of conic sections, has taught the theorist, the experimenter, the carpenter, and even the artist to find them everywhere, from the heavenly motions to the pose of a Venus. Applications are neither sought nor despised: They come of themselves. Classical geometry is so obviously, so egregiously applicable that no one asks what its applications are. Applications raise the human value of geometry and mechanics but do not change their natures.

The analogy to geometry should be drawn further. The early geometry laid out much thought on certain particular figures; triangle, square, circle, *etc.* Higher geometry requires a different method. It would be simply wasteful to go on in the same way with figure after figure. Rather, in modern geometry we learn properties of infinite classes of figures, such as polygons, conic sections, algebraic or analytic curves, and then take up more inclusive ideas such as metric, connection, and manifold. The geometric concepts are expressed directly in terms of *properties of invariance.*

So it is in rational mechanics. Not only to spare time but also to think simply and surely, we approach problems of material response in generality. Our physical experience with materials is summed up in statements of invariance. In Mr. NOLL's theory of materials, these are put directly into *mathematical language*. The theorist can no more be expected to tell the experimenter what constitutive functional to use than EUCLID tells the carpenter when to use a circle or when to use a square, or, in the terms preferred by physicists, "states the range of applicability and estimates the experimental error" of a triangle.

There are also persons who dismiss the modern work as being "axiomatics". While, unfortunately, we have not yet reached the point where a fully axiomatic treatment seems to be in order, the implication that things everyone knew already are just done over in a rigorous way is calumny. The foregoing lectures concern only results not known at all, or known but darkly, ten years ago. The modern work is *mathematical* and hence *strictly logical*, like classical geometry. Recall that after attempts scattered over three millenia and three continents, a proper set of axioms for Euclidean geometry was first obtained by HILBERT, in 1899. The older geometry, while not successfully axiomatic, was also not illogical, since it employed only logical methods. Rational mechanics, a logical or mathematical science, is in a pre-Hilbertian stage, and a completely axiomatic treatment lies some years ahead, but a start has been made.

In the first lecture I mentioned that one of the primitive elements of mechanics is the notion of a system of forces.

We are all used to mechanics as based upon a fundamental, primitive concept of force, but it was not always so. The word for force is older than all its measures. Since the earliest dynamical inquiries, motions and forces were associated, but what is the force of motion? What is the measure of force? The seventeenth century heard many debates on these questions, debates confusing to the debaters because they felt rather than specified the problems they sought to solve, and confusing to the modern reader because, with his slanted vocabulary, he is not even aware that there ever was something to argue about. The scientists of the seventeenth century had not been taught in school that when you have a problem of mechanics, the "forces" are "given", and you go ahead and set it up. When they saw a body moving on an inclined plane, or a mass swung at the end of a rope, or a top spinning on a point, or fluid split apart by the prow of a ship, they knew that "forces" were manifested, but how should these forces be measured? One was the "force of inertia", defined by KEPLER and mentioned by EULER, D'ALEMBERT and many others; this was equivalent to the modern word "inertia", a property of matter, but in the early literature it was an example of a "force", much as is "force of habit" in common speech today. For a long time there was an effort to define force in terms of motion: the weight times the velocity, the weight times the distance moved, the velocity times the distance moved, *etc.* The second lecture presented in Mr. TOUPIN's theory of polar-elastic materials the most complete and elegant example of success in a modern context, where use of an action principle renders the various kinds of force merely names for certain undetermined coefficients in a postulated extremal principle. Impressive as this trick may be for a system of this kind, it cannot be generalized to typical cases in mechanics, where dissipation is the rule, not the exception. Force is something more, something which must be added to the concepts of pure geometry and time. An *a priori* concept of force was, perhaps, NEWTON's main contribution, but in his writing it is rather illustrated than presented, explained, or defined explicitly, and at first only his own British disciples were converted to his way of looking at things. More than half a century after the publication of the *Principia*, D'ALEMBERT, expressing the view that what we call force is only a manifestation of motion, spoke of Newtonian forces as "obscure and metaphysical beings, capable of nothing but spreading darkness over a science clear by itself". It was through the dozens of examples worked out by DANIEL BERNOULLI and EULER that the fruitfulness of taking forces as quantities given *a priori* came gradually and grudgingly to be accepted. By common consent, nowadays, scientists and non-scientists rarely say anything concrete to each other; a scientist who uses the word "force" in its Newtonian meaning to musicians or poets will notice at once that understanding ceases, since to them "force" has only its pre-Newtonian senses even today.

While few scientists know that any non-Newtonian view of force ever existed, even fewer are able to tell us what a Newtonian force is. If we ask, we shall be told by a modern expert, "Force? Well, everybody knows about force intuitively. We all feel pushes and pulls, and we see experiments in which forces are applied. In the theory, we represent forces as vectors, and we assign them the dimensions of mass times acceleration, so they can go into NEW-TON's second law." This kind of talk would certainly not have impressed LEIBNIZ or D'ALEMBERT. It is accepted now for no better reason than that we are all used to it. We don't listen to the talk, really. Each of us in his own mind has a further associa-tion with forces that no book states. When we ask a mathematician what a real number is, he does not mumble, "Well, we all know about real numbers from our income-tax forms and from solving equations; they belong to an ordered field, you know, and we use them when we study calculus." Rather, he replies that a real number is an *undefined object* or *primitive*, described only by a definite set of *axioms*, axioms which he can state precisely and without hesitation. Now forces, too, are undefined objects; they are vectors, but saying that they are vectors no more characterizes forces than saying real numbers form an ordered algebraic field characterizes real numbers. For about a century we have known that there are many kinds of ordered algebraic fields, but that real numbers satisfy an axiom of completeness. Moreover, any field that does satisfy that axiom is indistinguishable from the real numbers. The first attempt to lay out the mathematical properties of a system of forces was by Mr. NOLL in 1960.

NOLL defines a *material universe* as a set of elements X called *particles*. It is given a structure by allowing smooth homeo-morphisms onto three-dimensional Euclidean space, the space of classical geometry, and by admitting a positive, finite measure defined over certain measurable subsets. The homeomorphisms are called the *configurations* of the material universe, and the measure, the *mass* of the measurable subsets.

A *body* $\mathscr{B}$ is a smooth, closed subset of the universe that has no closed, proper subsets endowed with all of its mass (unless that mass is zero). A body of mass zero is said to be *void*. Two bodies whose intersection is a void body are *distinct*.

Corresponding to each body $\mathscr{B}$ there is assumed to exist a distinct body, $\mathring{\mathscr{B}}$, such that the mass of the union of these two bodies is the mass of the universe. The body whose existence was just now postulated is called the *exterior* of the given body $\mathscr{B}$.

A *system of forces* for a material universe is a vector-valued function $f(\mathscr{B}, \mathscr{C})$ of pairs of bodies; the value of f is called the force exerted on the body $\mathscr{B}$ by the body $\mathscr{C}$. A system of forces is restricted by the following special properties, laid down as axioms:

1. For any fixed body $\mathscr{B}$, $f(\mathscr{C}, \overset{\circ}{\mathscr{B}})$ is a totally additive function defined over the sub-bodies $\mathscr{C}$ of $\mathscr{B}$. That is, the force exerted by the exterior of $\mathscr{B}$ on the various sub-bodies of $\mathscr{B}$ defines a vector measure over $\mathscr{B}$.

2. The same holds, conversely, for the force exerted on $\mathscr{B}$ by the sub-bodies of the exterior: For fixed $\mathscr{B}$, the function $f(\mathscr{B}, \mathscr{C})$ is a vector measure defined over all sub-bodies $\mathscr{C}$ of $\overset{\circ}{\mathscr{B}}$.

These properties formalize the idea that the parts of a body suffer forces independent of each other, and additively, from the parts of the environment. Hence we may define over the body an integration with respect to force.

The force exerted on $\mathscr{B}$ by the exterior of $\mathscr{B}$ is called the *total force* acting on $\mathscr{B}$.

The time-dependent system of forces, together with the motion of a body as defined in Lecture 1, constitutes a *dynamical process*. The scalar product of force by velocity, integrated over the body, defines the rate of working or *power* of the system of forces in the configuration at any given time.

Before going further we lay down the basic *Principle of Frame-Indifference* in a more general form: *All laws and definitions that hold in a dynamical process are the same for every observer.* Thus the principle of indifference asserts that all laws and definitions of mechanics must be invariant under time-dependent orthogonal transformations. In particular, body forces and contact forces are indifferent.

When the principle of indifference is applied to the definition of power, after some mathematical development the two foundation principles of mechanics follow as theorems valid for every dynamical process:

Theorem 1. *The resultant force acting on every body is zero.*

Theorem 2. *The resultant moment of force acting on every body is zero.*

In these statements, the force of inertia is not separated from other forces. The classical definition of the force of inertia rests on assigning a preferred status to a certain part of the universe, once and for all. Calling this part "the great system" and laying down a particular frame as a preferred frame, we eliminate consideration of bodies outside the great system if we *define* the force exerted by the universe exterior to the great system on any body in the great system as being the negative of the rate of change of momentum of the body with respect to the preferred frame. (In the traditional applications, the great system is the region of space out to the "fixed stars", which define the preferred frame, in which the force of inertia appears as the effect of the exterior, unspecified, unknown part of the universe (not necessarily "hidden masses") on the parts accessible to us.) Independently of any special interpretation, in the formal theory Theorems 1 and 2 then reduce to statements of the classical laws of balance:

Theorem 1*. *The resultant force exerted by the system on a body is the rate of change of momentum of the body (in the special frame).*

Theorem 2*. *The resultant moment of force exerted by the system on a body is the rate of change of the moment of momentum (in the special frame).*

These principles lie at the base of modern mechanical theories. Of course they contain as special and degenerate cases the principles of the old "analytical mechanics". In particular, they imply the so-called "Newtonian equations" in the stronger form, which requires that binary mutual forces in a system of mass points be central as well as equal and opposite.

In the same way as long-familiar geometrical objects, such as circles, squares, and tetrahedra, represent and abstract some aspects of the physical attributes of many objects in the world of the senses, so also in the mechanics of continua we seek to define *ideal materials* in such a way as to represent, abstractly and in part, the observed attributes of physical materials such as earth, water, air, and fire. These natural materials are, in the sense of chemistry, pure or compound or mixed, inert or reacting; in the sense of physics, solid or fluid or gaseous or of changing phase. All are *deformable*. In mechanics, it is their deformability, in gross, that we seek to abstract and reduce to law. Mechanics cannot, any more than geometry, exhaust the properties of the physical universe. No one, however, levies as a reproach against the geometric theory of the sphere the fact, true though it is, that a sphere of earth is a different thing from a sphere of fire. No one, at least no one among the informed, lays reproach on geometry because no sphere of water or air is a perfect geometric sphere, or because the geometry book has no special name for the figure of the latest rocket. Mechanics presumes geometry and hence is more special; since it attributes to a sphere *additional* properties beyond its purely geometric ones, the mechanics of spheres is not only more complicated and detailed but also, on the grounds of pure logic, necessarily *less widely applicable* than geometry. This, again, is no reproach; geometry is not despised because it is less widely applicable than topology. A more complicated theory, such as mechanics, is less likely to apply to any given case; when it does apply, it predicts *more* than any broader, less specific theory.

When we add to the geometric theory of space-time the general theory of systems of forces, we are in a position to set up the general theory of constitutive equations, which was outlined in the first lecture. Just as a relation among points or lines, or a sub-group of transformations, defines a special geometrical figure, a constitutive equation defines an *ideal material*.

As indicated in the third lecture, the next level of specialization introduces thermodynamic principles. In any domain, an organized, comprehensive theory can be formed only on the basis of success in special cases. While NOLL's theory of pure mechanics is the culmination of twenty years' intensive research on special materials, in rational thermodynamics since the time of GIBBS and DUHEM the first important paper is less than five years old, so we do not yet have experience sufficient for a definitive formulation, but clearly entropy is the new primitive concept that characterizes the subject, as was illustrated in the third lecture.

As mechanics is the science of motions and forces, so thermodynamics is the science of forces and entropy. What is entropy? Heads have split for a century trying to define entropy in terms of other things. Entropy, like force, is an *undefined object*, and if you try to define it, you will suffer the same fate as the force-definers of the seventeenth and eighteenth centuries: Either you will get something too special or you will run around in a circle.

The history of thermostatics has put a block between us and the understanding of entropy by denying that a substance called "heat" exists. If we were to start afresh, free of the bad associations originating in word-fights a century ago and devoutly embalmed in the physics texts today, we might call "heat" what now is called "entropy", thus bringing the concept as close to ordinary experience as are "force" and "mass". To say that heat is what goes into a body when it gets hotter or is deformed and usually goes out when it gets colder, and that the mean heat increases at least as fast as the mean increase of energy divided by temperature, seems to me as good as the explanation of force as being a push or pull in the amount of the mass times the acceleration produced, or as the explanation of mass as being the quantity of matter in a body.

Just as force can be defined kinematically in certain special and common systems, such as a pendulum, entropy can be, and is, defined thermally for certain special systems, such as a perfect gas. Just as force can never be measured directly — for all measurements of force presume some special mechanical constitutive equation such as that of the rigid body or the linear spring or the gravitational field — so entropy cannot be measured except indirectly through use of special thermal constitutive equations. While we can never define entropy except in cases where we do not need it, and while we cannot measure it directly, we can, in time, get used to it, as we have, despite the opposition of some of the greatest of all scientists, gotten used to an *a priori* concept of force. We are not yet sufficiently used to entropy to state, with any measure of agreement, what axioms it obeys.

These remarks, partly a summary of ideas illustrated in the preceding lectures, should make it clear that the method of modern natural philosophy is more or less that of any other branch of mathematics. Since it is the same method as was used by the great geometers of the seventeenth and eighteenth centuries, our colleagues in the professionalized natural sciences often look upon us as quaint antiquarians if not reactionaries from the radical right. Just as the university has changed from a center of learning to a social experience for the masses, so research, which began as a vocation and became a profession, has sunk to a trade if not a racket. We cannot fight the social university and mass-produced research. Both are useful — useful by definition, since they are paid, if badly. But we must not allow the social university to destroy learning, and the trade of research to take away our right and capacity to think. Society demands and pays for commercial art and canned music, but the employees in these industries do not hold themselves up as ideals toward which every painter and composer should strive. In contrast, the organized trade of science, not yet sufficiently distinct to boast its indifference to the old-fashioned individual ways, decries them as antiquated and evil, and seeks to strangle the vocation of science, dredging the public pocket as well as filling the public press with the triumphs of massive teams of "experts" lulled by the costly blink of binary numbers by the billion. Soon, perhaps, small children will skip from door to door, begging dimes for digits. No one will deny that the giant brains in obedience to teams of little ones can do things undreamt of by our fathers. What has not been shown is any change at all in the requirements for the kind of science the scientists of the past created. Granted that trade science can pour out in a day more tables of calculation and curves of experimental data than NEWTON could have inspected in a lifetime, I see no evidence that the kind of science NEWTON did, the science that has given, ultimately, the swarms of scientific ants the ground under their anthills, can be done in any other way than he did it. Today we all ride in trade-produced autos, but society does not jeer at athletes who run a race with their old-fashioned feet, just as runners ran in ancient Athens. The athletes and the artists are allowed to pursure their vocations, indeed, supported in them, in the midst of a beehive society. Likewise, a quiet corner must be found for the learned and for the creators, even in the modern university.

While not surrendering to trade science, the natural philosopher must not from misguided and suicidal snobbery attack it. Trade science is invincible, but it need not remain an enemy, for between trade science and science as a vocation, there is more misunderstanding than real conflict. Not so with religion science. By religion science I refer to the practices of those cults where the lingo and paraphernalia of science are used, not for inquiry, but for affirmation. The creative, intuitive act in science is replaced by violent faith in one or other devotional recipe. There are three ways that religion science can be recognized. First, it is endlessly repetitive. While true science, even trade science, joins fact to fact or theorem to theorem until, perhaps, the area it explores is understood and then largely abandoned as a field of study, the truths of a religion cannot be too often shouted or spoken or whispered or sung. While the greatest triumph of a scientist is to replace one whole corpus of concepts by another of his own making, in a religion the truth is already revealed, and of a successful composer of prayers no more is desired than skill in phrase and juxtaposition. When overwhelmed by great floods of "research" papers containing merely different wordings of the same prayer, we recognize the source as being a religion science.

Second, what is called "proof" in religion is not distinguished from revelation, and logic, to the extent it is used at all, is a device for erecting arpeggios upon a known hymn. So also when the literature of a science uses mathematical formulae, not as expressions of a chain of logic, but as ornaments for affirmations of faith, we recognize religion science.

Third, the basic tenets of a religion, and generally also the beliefs of its high priests, are true, true beyond question, true forever. They need only proper interpretation, proper exegesis, to cover every case. A scientifically phrased doctrine whose adherents claim this universality for it is likely to be religion science. In true science, there are mistakes. In mathematics, a mistake may be found by anyone — by a freshman, by an arrogant colleague from another department, by a failure in his profession who up to that time had never done anything correct in his life. A mathematician does not like to be wrong, but he may *be* wrong, and when he is, there is no doubt and no excuse. The truth is the truth, demonstrable, and independent of persons. In other human endeavors, truth is contingent. The politician, the lawyer, the physician, the general, the university official are all modest men, more modest than most mathematicians; they are the first to admit, in theory, that they are fallible, capable of error, perhaps even that in the (carefully unspecified) past they have not always been right; but today, here and now, they are *never* wrong, and nothing can make them wrong. The patient can die in agony, the army can be killed off to a man, the nation can be annexed and enslaved, the university can be overrun by a horde of students, but the error, if any there were, lies elsewhere than with those in command. When the terms of science are used to form and enforce a religion, the contingent infallibility of the boss, common and accepted in other areas of life, presents the danger of the devil preaching in a cassock. This danger is real, real not only to science but to the person of the natural philosopher today. Against the attrition of time, there is no bastion so strong as the truth, witnessed by myriad martyrs who neither fought nor died in vain; against the petty hierarchs of here and now, the truth is no more shield than a glass door. In the preceding lectures I have mentioned two areas of nature now occupied by religion science. Like islands first discovered by pirates, for no reason in their own soil they have never seen sound currency. Their overlords give to any unbiased scientific inquiry the welcome an honest man receives when he enters a racketeer's palace, and their closed-end corporations reject a gold piece as outmoded tender. A mathematician who has proved a theorem can safely leave that theorem to stand by itself, apart from personalities. Another mathematician may find an error, may better the proof, may criticize the theorem itself as being of little interest or insufficient generality, but he is unlikely to see any point in denouncing the author. Religion science treats criticism of its tenets as a personal attack on its priests and replies with personal abuse of the critic. Because I had published a sequence of specific and impersonal criticisms of a certain class of studies, I have just received from a professor whom I did not know to regard himself as hit a letter telling me I write "a pack of lies".

As LOCKIT warned us,

> "When you censure the Age,
> Be cautious and sage,
> Lest the Courtiers offended should be;
> If you mention Vice or Bribe,
> 'Tis so pat to all the Tribe;
> Each crys — That was levell'd at me."

Religion science is a greater danger than trade science, for religions burn heretics. It is conceivable that a great artist may be tolerated, even supported as a rare and precious bird, by a firm making beer posters. It is not conceivable that MICHELANGELO would have been allowed to paint CHRIST and the angels if he had been captured by the Caliph of Baghdad.

While trade science is a typical twentieth-century invention and hence, being social and profitable, is invincible, religion science is not. There have been religion sciences before. For example, in the early eighteenth century there was a rage among non-mathematical physicists for atoms of every sort, followed in the second half of the century by a rage for subtle fluids. These fads, while they never furnished a single proved theorem, easily explained everything in nature, and their priests sat high among the academicians of every land. Typically, the concepts chosen for veneration by these two cults, atoms and fluids, are in themselves in no way unworthy of science and have their firm places today. They were neither invented by the cult priests nor put to scientific use by them, merely worshipped, exploited, and used as weapons against the truth. These two cults are dead now. They were killed by the normal course of science, by the research of mathematicians like EULER and CAUCHY, who provided, in time, sound mathematical theories for the phenomena the atoms and the fluids were claimed to explain. To any real scientist, a body of theorems is clearly better than a heap of claims and sketches. The religion sciences of today will die, slowly, if left to the decision of scientists, but times have changed, and decisions, even decisions about science, are not made by the few arrogant eccentrics who are scientists in the old sense. Society reaches down into the farthest corners of the beehive, and society is the apotheosis of the common man. To the layman, be he a cog on an assembly line for toothbrushes or a true expert from a different compartment of science, religion science is indistinguishable from the truth. What the layman thought about science never mattered before. Now it is all that matters. In our day, the day of massive unanimity, the minuteness of science as a vocation, which in the seventeenth century was a guarantee of survival, is close to a guarantee of destruction. If religion sciences gain control of the production lines of trade science, the fate of true science will be like that of true history at the hands of the Inner Party in ORWELL'S *1984*.

The natural philosopher must stand firm against bullying by the cults, the professions, the teams, the government, the administrators, and the computing machines. His solitary role, just as it was in the seventeenth century, when society was no help but also no hindrance, is to think about nature and to understand it. This conservative quality is plain in the researches I have reported. Just as in the great age from HUYGENS to CAUCHY, an aspect of nature is faced, described, and developed in terms of the most powerful mathematics we can bring to bear. Some of the natural phenomena and much of the mathematics are new; it is the *method* that is traditional, identical with that of NEWTON and EULER and MAXWELL. We might transfer to a higher scene the apology of KELVIN & TAIT: "... where we may appear to have rashly and needlessly interfered with methods and systems of proof in the present day generally accepted, we take the position of Restorers, and not of Innovators."

If the method of modern natural philosophy is mathematical, so is its taste. While "good taste" is a term rarely used by natural scientists, it occurs frequently in conversations among mathematicians and sometimes even in research papers. Although the stranger to mathematics is likely to picture it as a factory for variants of the statement that $2+2=4$, leaving no room for imagination, fancy, or invention, and even engineers often seem to think that a mathematician will prove one thing as gladly as the next, so long as the proof be rigorous, in fact the choice of problem, hypothesis, and method of attack is at least as important in mathematics as is the choice of medium, subject, and pose in the arts. As mechanics, thermodynamics, and electromagnetism become mathematical sciences, style and taste grow more important in them.

It is tasteless to recommend one's own taste, but scarcely honest to recommend any other. Taste is learned best from experience and example, not from precepts, so I ought to say no more, except to warn against falling into the egalitarian fallacy of trying to make all persons identical.

Until the last year or two, the few odd-balls searching the foundations of "classical" physics were ignored as harmless. Now that "continuum mechanics" has become, overnight, a recognized "engineering science", they stand to be trampled by whole corps of the army of dulness, dense phalanxes of polynomial invariants and representations covering pages with hundreds of numbered coefficients, which are ground out by new-made "experts" much

as, not long ago, experts on "applied mechanics" found the
Fourier coefficients in series expansions for the stresses in an
elliptic plate pierced by three square holes. Lest I be thought to
exaggerate, I show as illustration[1] four consecutive pages from
the latest issue of an international journal devoted to "engineering
science". It is, I suspect, partly disgust with this sort of thing
that has caused two of the principal creators of the modern
theories, Mr. ERICKSEN and Mr. TOUPIN, in this past year to turn
their abilities back to the classical, linearized theory of elasticity
and find there new light on major old problems. In a famous
paper, which he reprinted as an appendix to the lectures he gave
at this university in 1884, Lord KELVIN spoke of "nineteenth
century clouds" over physics. There are clouds today over the
new natural philosophy; no signs from heaven, these are raised
by hoofs.

Foundation studies, when they are good, are very good; sprung
from experience, fancy, and art, they consolidate quickly within
the common culture of science at the very points, where, it seems,
they always naturally belonged and could never have escaped
discovery; but when they are bad, they are very bad, and the
bad money drives out the good. Natural philosophy, scarcely
reborn, must defend itself against attacks from without and cor-
ruption from within. On the outside, the high priests of religion
science threaten holy war against any apostle of reason. Within,
there are those who would make of natural philosophy one more
of the trade unions of science. Theory-creating is not a fit object
of mass research. Research as a trade, refusing to look in the mirror,
tries to preserve the illusion of independence if not leadership.
When terrain is explored by posses of volunteer householders, all
can be barons, but if we must have an army, not every man can
be a captain, let alone a general. Churning out theories, particularly those alleged to fit some experimental data, is as sterile
as the invention of special curves in geometry. Some special curves
or classes of curves, and some special theories or classes of theories,

---

[1] Since the pages flashed on the screen at the lecture could not be
reproduced in print without disclosing their author, and since deroga-
tion of persons would be the opposite of the effect intended by these
lectures, the reader is left the easy task of finding his own illustration
here.

deserve intensive and repeated study, but a randomly selected special theory, like a curve chosen by caprice, affords no more than a means of spending time and dirtying paper. A mathematical theory is empty if it does not go beyond a few postulates, definitions, and routine calculations. Theorems must be proved, theorems, good theorems. A great deal remains to be learned, for example, about the perfectly elastic solid and the simple fluid, concepts which, like conformal maps and analytic functions, will stay current in our culture as long as it lasts. The resulting theories are well defined and do not have to be drawn out by new researches on the foundations. They are ready for intensive study by scientists of many bents and at every level of competence.

Research has been overdone. By social command turning every science teacher into a science-making machine, we forget the reason why research is done in the first place. Research is not, in itself, a state of beatitude; research aims to discover something worth knowing. With admirable Liberalism, the social university has declared that every question any employee might ask is by definition a fit object of academic research; valorously defending its members against attacks from the unsympathetic outside, it frees them from any obligation to intellectual discipline; it brings the outside inside by abolishing the distinction between academic learning and any other activity that may be done without making money. An exhaustive, scrupulously catalogued collection of chewing-gum wrappers will soon seem at home in the National Gallery of Art. While once the title of "doctor" meant "teacher", now it is a mere certificate that study has stopped; while every cow college is now staffed solely by "leaders", we look in vain for those who follow them. If "research", too, has become a masonic grade, let us not forget that nature herself is unresponsive to the titles and honors we bestow upon each other. In natural philosophy, marvellous discoveries have been made in the last fifteen years. I hope that they will be learnt and understood. I have tried in these lectures to set an example, since, with a few minor exceptions, all the work I have reported here has been done by others. Rather than lecturing on my own research because it was I who did it, I have selected what seem to me the most important results in the new natural philosophy, studied them as best I could, and presented them to you because they are *worth learning*.

Closure

The matters here described are developed in full in the following treatises:

1. TRUESDELL, C., & W. NOLL: The non-linear field theories of mechanics, Flügge's Encyclopedia of Physics, vol. III, part 3, Berlin-Heidelberg-New York: Springer 1965.

2. TRUESDELL, C.: Ergodic theory in classical statistical mechanics (Lectures of May 23—31, 1960), p. 21—56 of Ergodic Theory. New York: Academic Press 1962.

Appendix

Text of the Chairman's Introduction* to the
Colloquium on the Foundations of Mechanics and Thermodynamics
held at the U.S. National Bureau of Standards, Washington,
October 21—23, 1959.

Fifty years ago classical mechanics, the science of the motion of masses as ordinarily encountered, seemed a finished discipline and hence dead. While it is still so presented to the student of physics, everyone here gathered knows that in the past two decades the entire foundation, if not always the aspect, of mechanics has changed. Indeed, we are here to look for common ground in facing what may be recognized as a crisis of mechanics. While recent researches have fenced within mechanics as definitely describable and predictable new ranges of phenomena, attempts to place a common solid foundation sufficient for them all have made us realize that now we know almost nothing.

The period of discovery was opened in 1940 by a beautiful paper of ECKART, which laid out part of a mathematical framework for describing the phenomenon of diffusion, among others, in a multiconstituent mixture. Diffusion is associated not only with mass transport but also with thermal differences and flow of heat; it is a fully thermomechanical phenomenon. ECKART remarked that the density of production of entropy is a bilinear form in certain variables, all of which vanish in equilibrium. Incorporating several earlier special theories, he proposed linear constitutive equations in which these factors are divided into two classes, any member of one being taken as a linear function of all the members of the other. Independently of this work, MEIXNER in 1941 proposed essentially the same theory but with two differences. First, he called attention to the support and interpretation given by the kinetic theory of monatomic gases to a

---

* (Previously published only in a German translation, "Zu den Grundlagen der Mechanik und Thermodynamik", Physikalische Blätter, Volume 16, pp. 512—517, 1960.)

special case of the phenomenological theory, and, second, he invoked a result inferred a decade earlier by ONSAGER on the basis of a statistical treatment of small fluctuations about equilibrium, namely, that the reversibility of the equations satisfied by the particles reflects itself in the possibility of choosing the members of the categories, now called "forces" and "fluxes", as expectations of a set of variables and their dynamic conjugates. Any possible linear relation among variables so selected, it was claimed, would necessarily have a symmetric or skew matrix, according to a rule of parity.

Much literature has arisen about this "irreversible thermodynamics". Many effects have been included; coefficients have been named, measured, and recombined in various ways; and several authoritative expositions have been published, but open questions remain.

1. If the "Onsager relations" are true, they must have not only a statistical but also a phenomenological meaning. Since the variables used in some of the major applications have never been shown to be expectations of conjugate quantities, phenomenological interpretation would be all the more useful. It seems difficult to interpret the "Onsager relations", however, until the "forces" and "fluxes" themselves are identified with particular physical quantities. When this is done, the need for the "Onsager relations" vanishes, since in a relation between well defined variables, if properly invariant, the nature of the coefficients is a matter for experiment or a particular molecular model. Perhaps the key lies in the matter of invariance. Is there a physically natural principle of invariance, expressed directly in terms of the phenomena, which applies? To decide this question, also, we must surely bring the "forces" and "fluxes" nearer to reality than are simply small x and large X. *In terms of the phenomena, what are the "forces" and "fluxes"?*

2. While mechanics is the science of motion, and motions certainly occur in the problems studied in irreversible thermodynamics, the theory follows the tradition of thermostatics rather than of dynamics in resting content with identifying, naming, and measuring causes, rather than determining motions. So far as I know, the literature does not contain a single worked-

out example of a motion except in cases reducible to subservience to the century-old theory of FICK. *Are the equations given in texts on irreversible thermodynamics sufficient to determine motions, or do new principles remain to be found?*

In 1945 appeared a paper by REINER in which, if we except a long prior formulation of finite elasticity, the first reasonably general and properly invariant non-linear theory of material behavior is proposed. Two years later was published the first of a series of remarkable papers by RIVLIN in which concrete problems, some of them of practical value, were solved explicitly and exactly, both for non-linearly viscous and for finitely elastic materials. If more general response, in any way variable in time, is to be included, an appropriate principle of invariance must be supplied so as to eliminate physically absurd behavior. The problem was recognized, and a correct solution was given, by OLDROYD in 1950. Better statements and more concise and comprehensive solutions have been given later; the principle, now called "material objectivity" or "material indifference", is a precise formulation of the physically natural idea that the response of the material is indifferent to the observer.

The results achieved by the non-linear theories of materials are elegant, unequivocal, and rigorous, but limited in scope. Closely bound in concept to older ideas in rational mechanics, they render explicit the reaction of mass but neglect the reaction of energy except when it can be set aside as an effect determined by, rather than contributing to, the cause of motion. Typically, the specific problems solved refer to incompressible materials. To obtain a determinate formulation for compressible media, as is known even for ordinary gas dynamics, not only dynamic but also energetic constitutive equations are required. While in inviscid gas flow the equations of state for equilibrium are assumed to hold locally, from special results in the kinetic theory of gases or from studies of irreversible thermodynamics we see that no such simple expedient will yield an adequate description of the energetics of large deformation of non-linear materials. As a guide in formulating thermo-mechanical or more general theories, in 1949 I proposed a requirement which has since been clarified and named "the principle of equipresence": The independent variables in all constitutive equations should be the same. In terms of a molecular

model, this principle corresponds to the fact all that phenomenological quantities, such as stress and heat flux, are gross mani-
festations of the same laws, namely, the laws of motion of the
corpuscles. It is similar to the "Onsager relations" in that when
it has once been used properly and fully, it will have been incorporated in the general equations and hence exhausted. To apply it,
however, we have to know what the independent variables are.
The student of rational mechanics sees the flux of energy as the
analogue of the stress tensor in expressing a superficial equivalent
to the action of the exterior upon the interior; these are the
dependent variables of thermo-mechanics. But on what do they
depend? Surely, upon the deformation history, but what independent variable describes the energetic aspect of the problem?
If FOURIER's law is to be included, the temperature gradient must
be one of the energetic variables, but if we turn for guidance to
thermodynamics, we find the temperature itself playing the role
of the driving force. No simple expedient seems to serve to put
all these ideas together, and we find ourselves faced with a major
open question:

3. *What are the independent variables of thermo-mechanics?*

However, in trying to follow through the thoughts embodied
in primitive or degenerate form in the basic equations of gas
dynamics, we quickly see that the caloric equation of state,
itself, is the weakest link. Does it express a physical principle, or
is it a constitutive equation, like a law of viscosity or conductivity?
The kinetic theory of gases suggests, and I think the deeper students of the subject now adopt, the latter view, with the corollary
that classical thermodynamics based upon an equation of state is
a subsidiary and approximate discipline. For this reason, researches
in rational mechanics have not been able to derive any help from
"irreversible thermodynamics", which assumes a simple answer to
the very question they face, namely:

4. *What is the basic principle of entropy production for large
irreversible deformation?*

We must seek, then, a true thermodynamics as a class of
constitutive equations, different in concept and more embracing in
principle than the classical thermostatics, which may be expected
to emerge, along with the linear thermodynamics of irreversible
processes, as a limit case for slow motions.

That classical statistical mechanics had early divided itself into two disjoint disciplines, the theory of transport processes in dilute monatomic gases and the theory of equilibrium or of small perturbations about equilibrium in more complicated systems, was not without reason. The Maxwell-Boltzmann integro-differential equation of the kinetic theory was discovered by semi-statistical arguments of plausibility. While little doubt was expressed regarding the result, the insufficiency of its foundation was proved by the want of any corresponding theory of liquids, for the methods of inference were plainly inadequate to get anything better.

In order to approximate, it is necessary to know what is being approximated; a scheme of approximation which does not contain, in principle, a systematic method of improvement is no more than a guess. Problems of this kind in statistical mechanics are too vast for fruitful discussion unless narrowed as follows:

5. *Within general statistical mechanics, for what circumstances, if any, is the Maxwell-Boltzmann equation a valid approximation?*

6. *What equations govern the temporal evolution of molecular distribution functions in imperfect gases and liquids?*

These problems were recognized and attacked in memoirs published in 1946 by BOGOLUBOV, KIRKWOOD, and BORN & GREEN. These authors set up formal approximative processes in terms of which a definite formulation, capable of subsequent proof or disproof, is possible. In particular, the previously known process for calculating the coefficients in a series expansion of the general equation of state for equilibrium is amplified so as to yield, at each stage, a corresponding modification of the law of change in time for disequilibrated distributions. A new field of research, the general theory of transport processes, developed, and many concrete results of real interest have been derived. It is not in underestimation of these, but rather in appreciation of the great difficulty of a genuinely statistical treatment, that I say that the two major questions remain open. The researches of the past decade have made the subject more rigorous than it was; there is reason to hope that further study along the same lines may resolve the basic questions with the rigor that brings certainty.

Only after these questions are answered can we face the problem where the statistical and phenomenological approaches come together:

7. What are the hydrodynamical laws according to the kinetic theory?

Though rarely so attacked, this question is a purely mathematical one. As a sample of its slipperiness I mention only that in the classical kinetic theory it has been shown possible, by a specific example, that a relation between stress and rate of shearing which is correct to within an error which vanishes rapidly with time may lead, if substituted into the equations of motion, to formulae for the gaseous stresses which grow with time to have an infinite proportional error. Nothing could be more illuminating for theories of gross material response than a reliable relevant example from the statistical mechanics of even a highly simplified special model, but nothing can be more misleading than a specious argument drawn from unrigorous statistical ideas. The open problems of the statistical mechanics of transport processes still lie at the fundamental level; comparison with phenomenological theories or experiments on large irreversible deformations would be premature.

A major aspect of classical physics remains unmentioned: electromagnetism. The first rational theory of a large electromechanical interaction was published by TOUPIN in 1956. The properties of invariance enjoyed by the classical electromagnetic equations are different from those of the equations of classical mechanics. Since the principle of material indifference asserts that the definition of a material is to be invariant under more general circumstances than are the laws of mechanics, the diverse nature of electromagnetic invariance would not in itself be reason to alter the requirements for constitutive equations. However, it is the classical concept of "observer", based upon the notion of rigid frame, that leads to the principle of material indifference, and in electromagnetism a different concept of "observer" is more natural. When two different approaches, each successful in its own domain, seem to meet in conflict, both may well be too special, and perhaps it is right to abandon the more restrictive aspects of each. Here there has been a revival of the program of KOTTLER, who, after showing that the electromagnetic conservation principles may be formulated invariantly in a fully general space-time, making no commitment as regards geometry, suggested that properly understood physical laws ought always to be stated without geometric presumptions and should in fact determine the structure of space-time. By introduction of a sufficiently general kind of differential invariant, depending upon two events rather than only one, mechanics and electromagnetism may indeed be combined. In order to formulate a proper guiding principle for definitions of materials within this unified theory, we are then faced with the question:

8. *What is an observer?*

So far, we have noticed problems which arise in seeking to combine mechanics with thermodynamics *or* with electromagnetism. In the world-invariant scheme already mentioned, energetics is naturally incorporated, with internal energy and flux of energy entering a general conservation law along with velocity and stress. Temperature and entropy, however, are not mentioned. If we seek a rational union of electromagnetic, thermal, and mechanical concepts, we face again the questions already stated in connection with mechanics, namely, what are the independent variables and what is the law of entropy production?

As sum of our knowledge in the field we are met to consider, I may have seemed to write "zero". This is not so. The speakers will doubtless include details of some concrete achievements; only by these have we learned enough to delimit our ignorance. In fact, the gain is greater. Physics is not the only science in which intuition guides. There is also mathematical intuition. "Intuition" is a term introduced so as to glorify or deplore the use of mentally organized experience to extrapolate or to generalize. The mind, whether scientific or non-scientific, symbolizes experience by what the scientist sometimes calls "models" but the well balanced, normal citizen calls "over-simplifications" or "exaggerations" if recognized by him because used by another citizen to support a conclusion in which he does not agree. The models familiar in different walks of life are different from one another, and correspondingly different rules are allowed for constructing and displaying them. When "the mathematician", "the physicist", and "the chemist" study the same phenomena, as once in a while occurs, their processes of building models are often different even if, as still more rarely occurs, their results are more or less the same. The consequent reluctance to accept each other's models is sharpened by the current spavined programs of graduate study, for the man who was trained in a department of physics or chemistry has been encouraged to overestimate the role of experimental experience in suggesting and supporting his arbitrary models and to underestimate the strength of his own mind in forming them, while the man who has been pronounced fit by a department of mathematics is likely to be more than confident of his intellectual independence while blind to the illumination of physical rather than mental experience. With this understanding of the nature of intuition, it is fair to say that much of the work done in the past twenty years has served mainly to build up intuition. Several of the conceptual blocks faced a decade ago have been dissolved; when we look back at the papers we ourselves wrote then, they seem to creak and groan in getting to halves of what are now to us obvious principles or results, being incorporated into our intuition of the mechanics of material behavior.

Our speakers are of backgrounds at least as diverse as the terms "mathematics, physics, and chemistry" suggest. They have been led to attack the embracing problem we are here to discuss through encountering specific cases; most of them have published statistical as well as phenomenological researches. Neither they nor we are here to snarl over the superiority of one or another corner or to gain converts. Each wishes to proffer his suit in his own style, nor are we to expect all the major questions of mechanics to be settled by majority vote at the end of the session. Rather, I should say, we are here to share our intuitions.

Comment appended in February, 1965

Answers to five of the questions of 1959 may be given now, as the foregoing lectures show.

Question 1. In terms of the phenomena, what are the "forces" and the "fluxes"? Answer (Lecture III): *irrelevant question*.

Question 2. Are the equations given in texts on irreversible thermodynamics sufficient to determine motions, or do new principles remain to be found? Answer (Lecture III): *No and no*.

Question 3. What are the independent variables of thermo-mechanics? Answer (Lecture III): *the histories of the deformation and temperature fields.*

Question 4. What is the basic principle of entropy production for large irreversible deformation? Answer (Lecture III): *the classical Clausius-Duhem inequality, applied to constitutive equations.*

Question 8. What is an observer? Answer (mentioned in Lecture IV): *a non-sentient invariant.*

Question 5—7 concern the far more difficult subject of the statistical mechanics of highly deformable systems. With the brilliant exception of the work of H. GRAD, the vast literature shows little evidence of facing the facts. Here lies a major challenge to the natural philosopher today.

Druck der Universitätsdruckerei H. Stürtz AG., Würzburg